T0141888

Combinatorial Optimization and Graph Algorithms

Takuro Fukunaga · Ken-ichi Kawarabayashi
Editors

Combinatorial Optimization and Graph Algorithms

Communications of NII Shonan Meetings

 Springer

Editors
Takuro Fukunaga
National Institute of Informatics
Tokyo
Japan

Ken-ichi Kawarabayashi
National Institute of Informatics
Tokyo
Japan

ISBN 978-981-13-5581-3   ISBN 978-981-10-6147-9   (eBook)
DOI 10.1007/978-981-10-6147-9

Printed on acid-free paper

This Springer imprint is published by Springer Nature
The registered company is Springer Nature Singapore Pte Ltd.
The registered company address is: 152 Beach Road, #21-01/04 Gateway East, Singapore 189721, Singapore

Preface

In a very short time since the concept of "P" and "NP" was introduced, algorithmic aspect of mathematics and computer science has quickly gained explosion of interest from the research community. Indeed, the problem of whether $P = NP$ (or $P \neq NP$) is one of the most outstanding open problems in all of mathematics. A wider class of problems from mathematics, computer science, and perhaps operations research has been known to be NP-complete, and even today, the collection of NP-complete problems grows almost every few hours.

In order to understand NP-hard problems, the research community comes to have a fairly good understanding of combinatorial structures in the past 40 years. In particular, the algorithmic aspects of fundamental problems in graph theory (i.e., graph algorithm) and in optimization (i.e., combinatorial optimization) are two of the main areas that the research community have some deep understandings.

Graph algorithm is an area in computer science that tries to design an efficient algorithm for networks. Today, networks (i.e., graphs) are ubiquitous in today's world; the World Wide Web, online social networks, and search and query-click logs can lead to a graph that consists of vertices and edges. Such networks are growing so fast that it is essential to design algorithms to work for such large networks. So the research community works on theoretical problems, as well as on some practical problems that implementation of an algorithm for large networks.

Combinatorial optimization is an intersection of operations research and mathematics, especially discrete mathematics, which deals with some new questions and new problems from practice. It is trying to find an optimum object from a finite set of objects. In order to tackle these problems, we require the development and combination of ideas and techniques from different mathematical areas including complexity theory, algorithm theory, matroids, as well as graph theory, and combinatorics, convex and nonlinear optimization, discrete and convex geometry.

As mentioned before, combinatorial optimization and graph algorithms are the areas that have attracted a lot of attentions right now. This can be also witnessed by the fact that there are many NII Shonan meetings in this area, including graph algorithms and combinatorial optimization in February 2011 (which is the first Shonan meetings) and combinatorial optimization in April 2016. Both meetings have attracted many top researchers in combinatorial optimization and graph algorithms, as well as many young brilliant researchers.

This volume is based on several Shonan meetings. We focus on theoretical aspects of combinatorial optimization and graph algorithms, and we have asked several participants to extend their lectures into written surveys and contribute to this volume. We are fortunate to have five renowned researchers in this volume. It covers variety of topics, including network designs, discrete convex analysis, facility location and clustering problems, matching game and parameterized complexity.

This volume consists of five chapters. Let us look at each chapter more closely.

Facility location is one of the central topics in the combinatorial optimization. In this problem, we are given a set of facilities and clients, and the problem demands choosing subset of facilities to open and assigning every client to one of the open facilities so that the opening costs of facilities and the distance from each client to its assigned facility are minimized. The facility location problem can be seen as a problem of clustering a given set of clients. By changing the problem setting slightly, we can also define the k-median and k-center problems, which are extensively studied problems in this field. Since these problems are all NP-hard, approximation algorithms have been considered mainly. Indeed, studies on the facility location and its variants lead to the discovery of many algorithmic methodologies so far. In "Recent Developments in Approximation Algorithms for Facility Location and Clustering Problems" by An and Svensson, they survey recent progress on approximation algorithms for these problems. In particular, they focus on two algorithmic methodologies—local search and linear programming based methods.

In "Graph Stabilization: A Survey," Chandrasekaran introduces graph stabilization, which is an optimization problem motivated by the cooperative matching game. In the cooperative game, a distribution of the profit to players possessing some desirable property is called a core allocation. Stable graphs are those graphs for which the matching game has non-empty core. The graph stabilization seeks a smallest modification of a given graph to make it stable. Besides the motivation from the game theory, the graph stabilization is interesting because it has a connection with matching theory, integer programming, and linear programming. It has received considerable attention recently, and much progress has been made in the research on it. Chandrasekaran surveys this progress.

Network design problems are optimization problems of constructing a small-weight network and include many fundamental combinatorial optimization problems such as the spanning tree, Steiner tree, and Steiner forest problems. In "Spider Covering Algorithms for Network Design Problems," Fukunaga surveys spider covering algorithms, which are key technique for solving various network design problems. Spider covering algorithms rely on the idea used in the analysis of the well-known greedy algorithm for the set cover problem. To extend this idea to network design problems, a type of graph called a spider is used in spider covering algorithms. They construct a network by repeatedly choosing low-density spiders. The framework of these algorithms are originally invented for designing an approximation algorithm for the node-weighted Steiner tree problem and has been sophisticated in a series of studies on numerous network design problems. Fukunaga introduces the basic idea of the spider covering algorithms.

Discrete convex analysis is a theory of convex functions on discrete structures and has been developed in the last 20 years. In "Discrete Convex Functions on Graphs and Their Algorithmic Applications," Hirai is a pioneer to explore the theory of discrete convex analysis on graph structures. His research is motivated by combinatorial dualities in multiflow problems and the complexity classification of facility location problems on graphs. In his chapter, the recent development of his theory and algorithmic applications in combinatorial optimization problems is discussed.

The problem of finding an assignment of authorized users to tasks in a workflow in such a way that all business rules are satisfied has been widely studied in recent years. What has come to be known as the workflow satisfiability problem is known to be hard, yet it is important to find algorithms that can solve the problem as efficiently as possible, because it may be necessary to solve the problem multiple times for the same instance of a workflow. Hence, the most recent work in this area has focused on finding algorithms to solve the problems in the framework of "parameterized complexity," which deal with problems in which some parameter k is fixed (we call "Fixed Parameter Tractable"). In "Parameterized Complexity of the Workflow Satisfiability Problem" by D. Cohen et al., they summarize our recent results.

We are very grateful to all our contributors and the editors of Springer Nature. Finally, we would like to thank the organizers of the corresponding Shonan meetings and their participants. Special thanks also go to NII for organizing the Shonan workshops, which now becomes the Asian "Dagstuhl."

Tokyo, Japan
July 2017

Takuro Fukunaga
Ken-ichi Kawarabayashi

Contents

Recent Developments in Approximation Algorithms for Facility Location and Clustering Problems

Hyung-Chan An and Ola Svensson

Abstract We survey approximation algorithms for facility location and clustering problems, focusing on the recent developments. In particular, we review two algorithmic methodologies that have successfully lead to the current best approximation guarantees known: local search and linear programming based methods.

1 Introduction

The *facility location problem* is one of the most fundamental problems widely studied in operations research and theoretical computer science: see, e.g., Balinski [7], Kuehn and Hamburger [32], Manne [37], and Stollsteimer [40]. In this problem, we are given a set of facilities $\mathcal{F}$ and clients $\mathcal{D}$, along with a distance function $c : \mathcal{F} \times \mathcal{D} \to \mathbb{R}_+$ and opening costs $f_i \in \mathbb{R}_+$ associated with each facility $i \in \mathcal{F}$. Given this input, the goal of the problem is to choose a subset of facilities to "open" and to assign every client to one of the open facilities while minimizing the solution cost, where the cost is defined as the sum of the opening costs of the chosen facilities and the distance from each client to its assigned facility.

The facility location problem, in its essence, is a clustering problem: we aim at finding a low-cost partitioning of the given set of clients into clusters centered around open facilities. While the facility location problem penalizes trivial clusterings by imposing opening costs, most of the other versions of clustering problems specify a hard bound on the number of clusters instead. Along with this hard bound, different objective functions give rise to different clustering problems. The *k-median problem*, for example, minimizes the total distance between every client and its assigned facility, subject to the constraint that at most k facilities can be opened (but opening

H.-C. An (✉)
Department of Computer Science, Yonsei University, Seoul 03722, Korea
e-mail: hyung-chan.an@yonsei.ac.kr

O. Svensson
School of Computer and Communication Sciences, École Polytechnique
Fédérale de Lausanne, 1015 Lausanne, Switzerland
e-mail: ola.svensson@epfl.ch

© Springer Nature Singapore Pte Ltd. 2017
T. Fukunaga and K. Kawarabayashi (eds.), *Combinatorial Optimization and Graph Algorithms*, DOI 10.1007/978-981-10-6147-9_1

costs do not exist). The facility location problem in fact is closely related to the k-median problem as the former can be considered as a Lagrangian relaxation of the latter. The k-*center problem* is a bottleneck optimization variant, where we minimize the longest assignment distance rather than the total. Finally, the objective function of the k-*means problem* is the sum of the squared assignment distances. As we will see soon, techniques devised in one of these problems often extend to others, letting us to consider these problems a family.

All of these problems have been intensively studied to lead to the discovery of many algorithmic methodologies; however, one of the common characteristics of these problems, namely that an open facility can be assigned an unlimited number of clients, can become an issue when we apply these problems to practice. In order to address this difficulty, the *capacitated facility location problem* was proposed. In this problem, each facility $i \in \mathcal{F}$ is associated with a maximum number of clients $U_i \in \mathbb{Z}_+$ it can be assigned, called *capacity*. This practically motivated variant, naturally, also exists for the other (hard-bounded) clustering problems. Unfortunately, these capacitated variants seem to make fundamental difference in terms of the problems' difficulty: many algorithmic techniques devised in the context of uncapacitated problems did not extend to the capacitated ones.

Most of the problems we have discussed above are NP-hard, and we therefore do not expect that efficient algorithms exist for these problems. Faced with NP-hard problems, we often overcome the difficulty by pursuing an approximation algorithm which trades off the solution quality for fast computation. Formally, a ρ-approximation algorithm (for a minimization problem) is defined as an algorithm that runs in polynomial time to produce a solution whose cost is no worse than ρ times the optimum. As the facility location problem in its most general form is SETCOVER-hard, we often focus on the case where c is metric, i.e., c satisfies the triangle inequality. Moreover, in the k-means problem, we usually consider the case where the clients lie in an Euclidean space and facilities can be chosen from the same continuous space.

In this article, we will survey approximation algorithms for facility location and other clustering problems, putting emphasis on recent developments. In particular, we will discuss two main algorithm techniques that have been successful in devising such algorithms: local search and linear programming (LP) based methods.

2 Local Search Algorithms

In the local search paradigm, we define an adjacency relation between feasible solutions to the given problem, where this adjacency usually corresponds to a small change in the solution. The algorithm is then to start from an (arbitrary) feasible solution and to repeatedly walk to an adjacent improved solution until we arrive at a local optimum. The analysis, accordingly, is in two parts: first, we establish that any local optimum is a globally good approximate solution; second, we show that we arrive at a local optimum within polynomial time. Sometimes these two parts are obtained by slightly modifying the algorithm so that it starts from a carefully chosen feasible solution and/or it finds a local near-optimum.

While a solution to the facility location problem (and to k-median, k-center, and k-means) is described by a subset of open facilities along with an assignment of the clients, without loss of generality we can assume that a solution is fully characterized by the subset of open facilities only: we can easily determine the optimal assignment subject to a given set of open facilities by assigning each client to its closest open facility. Local search algorithms for facility location and other clustering problems thereby consider feasible solutions simply as subsets of facilities.

2.1   Uncapacitated Facility Location

A local search algorithm for the facility location problem was first proposed by Kuehn and Hamburger [32] in 1963. While it was empirically known that this algorithm performs well [22, 41], it was only in 1998 when the first analysis that shows this algorithm is an $O(1)$-approximation algorithm was given [30]. This local search algorithm defines two solutions $S, T \subseteq \mathcal{F}$ to be adjacent if and only if $|S \setminus T| \leq 1$ and $|T \setminus S| \leq 1$. That is, for a given solution S, its adjacent solutions can be obtained by one of the following three operations: adding a facility, removing a facility, and swapping a facility. For some $S \subseteq \mathcal{F}$, let $C(S)$ denote its cost, i.e., the sum of the opening costs of the facilities in S and the optimal assignment distance of $\mathcal{D}$ to S. Following is the key lemma to show that this algorithm is a $(9 + \varepsilon)$-approximation algorithm[1] for any $\varepsilon > 0$:

Lemma 1   (Korupolu et al. [30]) *Let S^* denote an optimal solution. If $S \subseteq \mathcal{F}$ satisfies $C(S) > (9 + \varepsilon) \cdot C(S^*)$, there exists $T \subseteq \mathcal{F}$ adjacent to S such that $C(S) - C(T) \geq \frac{1}{\mathrm{poly}(n)} C(S)$.*

Proof (sketch) If a large portion of $C(S)$ is from the *connection cost*, i.e., the sum of the distances from each client to its assigned facility, we can show that adding a facility significantly improves the solution cost. Since $C(S)$ is dominated by the connection cost, under an appropriate choice of constants (used to define what a "large portion" means),[2] we have that the connection cost of the current solution alone is already much larger than $C(S^*)$. (Recall that $C(S) > (9 + \varepsilon) \cdot C(S^*)$.) Consider the process of simultaneously adding every facility in S^* to the current solution. This process improves the solution cost in spite of the added facility costs, and we can split and attribute this total improvement to the facilities in S^*. From an averaging argument, we can then show that adding one of the facilities in S^* gives an improvement of at least $\frac{1}{\mathrm{poly}(n)} C(S)$.

Now consider the other case where $C(S)$ is dominated by the opening cost. We will show that removing or swapping a facility significantly improves the solution

---

[1] This factor was improved to $(5 + \varepsilon)$ in the journal version [31] of the same paper.

[2] This choice of constant must be large enough to offset the choice of 9 in the lemma's statement.

cost in this case. For each facility i, we can compute its "per facility" connection cost by taking the sum of the distances between i and the clients assigned to it. We then identify a subset of facilities $\bar{S}$ whose per facility connection cost is relatively small compared to its opening cost. Since the entire solution cost $C(S)$ is dominated by the opening cost, a majority of facilities in S will be in $\bar{S}$.[3] We further restrict $\bar{S}$ to $\hat{S}$ by selecting the facilities whose opening cost is larger than (a constant fraction of) the average opening cost of $\bar{S}$. Note that $\hat{S}$ again, by an averaging argument, constitutes a major part of $\bar{S}$ and, in turn, of S.

We then draw a ball B_i around each facility $i \in \hat{S}$ with radius proportional to its "per demand opening cost", given by the opening cost divided by the number of clients assigned to it. Now we consider two cases: if two of these balls intersect, this means the centers of the two balls are close to each other. Thus, if we remove one of these two facilities and reassign its clients to the remaining facility, we will lose little in connection cost but earn a lot in opening cost. We have that removing a facility in $\hat{S}$ gives an improvement of at least $\frac{1}{\text{poly}(n)} C(S)$ in this case.

The other case we consider is when these balls are disjoint. In this case, we can show that there exists at least one ball, say B_i, that contains a *cheap* facility $i^* \in S^*$. Swapping i with i^* in this case leads to an improvement: since i^* is in the ball, we will lose little in connection cost; on the other hand, we earn a lot in opening cost since i^* is cheap. It now remains to argue that such a ball always exists; this can be done by showing that $C(S^*)$ would otherwise be as large as the opening cost of S, contradicting our assumption $C(S) > (9 + \varepsilon) \cdot C(S^*)$. Suppose that every ball either contains only an expensive facility in S^* or none of the facilities in S^*. If a ball B_i does not contain a facility in S^*, this means that, in the optimal solution, every client in B_i must be assigned to a facility outside the ball, implying that their assignment distance is larger than the radius of B_i.[4] This, in turn, implies that the *total* connection cost of these clients in the optimal solution is comparable to the opening cost of i.[5] (Recall that $i \in \bar{S}$ and therefore its opening cost is comparable to its per facility connection cost.) On the other hand, if B_i contains an expensive facility $i^* \in S^*$, we can charge the opening cost of i to i^* because i^* is expensive. Observe that each facility in S^* can be charged at most once because the balls are disjoint. Summing these bounds up, we can show that (part of) $C(S^*)$ is already as large as the opening cost of S. We therefore have that swapping a facility in this case gives an improvement of at least $\frac{1}{\text{poly}(n)} C(S)$. $\square$

This lemma yields the following theorem.

Theorem 1 (Korupolu et al. [30]) *For all $\varepsilon > 0$, there exists a $(9 + \varepsilon)$-approximation algorithm for the facility location problem.*

---

[3]We have that the opening cost of $\bar{S}$ is greater than a constant fraction of the opening cost of S; moreover, this inequality has a small slack that is proportional to the per facility connection cost of $\bar{S}$.

[4]This, of course, is true only for those clients that are near i, the center of the ball.

[5]The comparison has a slight error (see Note 4), which is charged against the slack described in Note 3. This balancing, along with the choice of constant in Note 2, determines the overall approximation ratio of 9.

Proof From Lemma 1, we can show that a polynomial number of local improvement reduces the solution cost by a constant factor, unless we prematurely arrive at a $(9 + \varepsilon)$-approximate solution. Since the cost of any arbitrary feasible solution can only be exponentially away from the optimum, it is clear that the local search algorithm arrives at a $(9 + \varepsilon)$-approximate solution within a polynomial number of iterations. On the other hand, each iteration of the algorithm involves examining $O(n^2)$ adjacent solutions and computing their costs in polynomial time, showing that the overall running time is bounded by a polynomial in the input size.   □

The current best approximation ratio based on local search is $1 + \sqrt{2} + \varepsilon$ due to Charikar and Guha [14], obtained by using a more sophisticated local search algorithm. Their algorithm provides a better performance guarantee for the opening cost portion than the connection cost; we accordingly can give as input a scaled-up version of the opening costs to this algorithm, achieving the overall performance of $1 + \sqrt{2} + \varepsilon$. This technique is called *cost scaling*, and has been used in conjunction with LP-based algorithms as well.

Theorem 2 (Charikar and Guha [14]) *For all $\varepsilon > 0$, there exists a $(1 + \sqrt{2} + \varepsilon)$-approximation algorithm for the facility location problem.*

In what follows, we describe the use of local search first for capacitated facility location and then for k-median and k-means.

2.2 Capacitated Facility Location

The best approximation algorithms known for the capacitated facility location problem are based on the local search paradigm. The first constant factor approximation algorithm was obtained in the special case of uniform capacities (all capacities being equal) by Korupolu et al. [31] who analyzed the simple local search heuristic proposed by Kuehn and Hamburger [32] previously described for the uncapacited version. Their analysis was then improved by Chudak and Williamson [18], and the current best approximation algorithm for this special case is a local search by Aggarwal et al. [2]:

Theorem 3 (Aggarwal et al. [2]) *For all $\varepsilon > 0$, there exists a $(3 + \varepsilon)$-approximation algorithm for the capacitated facility location problem in the special case of uniform capacities.*

For the general problem (i.e., non-uniform capacities), Pál et al. [38] gave the first $O(1)$-approximation algorithm. Since then more and more sophisticated local search heuristics have been proposed, the current best being a recent local search by Bansal et al. [8]:

Theorem 4 (Bansal et al. [8]) *There exists a 5-approximation algorithm for the capacitated facility location problem.*

We remark that the best local search algorithms known for the capacitated facility location problem are significantly more sophisticated than those explained for the uncapacitated version. We refer the interested reader to Pál et al. [38] and Bansal et al. [8].

2.3   k-median and k-means

As k-median and k-means only allow us to open at most k facilities, the natural local search for these problems only allow the swapping operation so that the number of open facilities is preserved. In other words, starting with any solution that opens k facilities, they repeatedly do swaps (i.e., open a new facility while closing a previously opened one) that improve the objective function. This basic local search can also naturally be extended to that of allowing p-swaps instead of only simple swaps. In a p-swap, we close p facilities and open p facilities. Note that for any constant p, we can check whether there is an improving p-swap in polynomial time. By selecting p as a function of ε, the following results were obtained for k-median and k-means.

Theorem 5 (Arya et al. [6]) *For all $\varepsilon > 0$, there exists a constant p such that the local search algorithm with p-swaps is a $(3 + \varepsilon)$-approximation algorithm for the k-median problem.*

Recall that the typically studied version of the k-means problem is the one where the clients and facilities lie in a continuous Euclidean space. In what follows, we also make the same assumption.

Theorem 6 (Kanungo et al. [29]) *For all $\varepsilon > 0$, there exists a constant p such that the local search algorithm with p-swaps is a $(9 + \varepsilon)$-approximation algorithm for the k-means problem.*

We remark that these papers also show that their analyses are tight. For relevant special cases, on the other hand, there has been very interesting recent progress. Independently by Cohen-Addad et al. [19] and Friggstad et al. [23], the basic local search with p-swaps was shown to give a polynomial-time approximation scheme (PTAS) for the Euclidean metric of fixed dimensions.

Theorem 7 (Cohen-Addad et al. [19] and Friggstad et al. [23]) *Consider the k-means problem in the Euclidean metric space of fixed dimension d. For all $\varepsilon > 0$, there exists an integer p that depends on d and ε such that the local search algorithm with p-swaps is a $(1 + \varepsilon)$-approximation algorithm.*

The two papers also give more general conditions under which k-means admit a PTAS: Friggstad et al. [23] show that the same result can be obtained for the more general doubling metrics and Cohen-Addad et al. [19] show an analogous result for graph metrics of graphs with forbidden minors such as planar graphs for example.

3 LP-Based Algorithms

Another important class of algorithmic techniques that has proven to be useful in attacking facility location and clustering problems is LP-based methods. Given a combinatorial optimization problem, we can obtain its LP formulation by encoding a feasible solution as a set of variables and writing the feasibility constraints as linear inequalities of these variables. This LP can be solved in polynomial time to obtain a bound on the optimal solution value, and furthermore, the LP solution itself often roughly reflects the structure of the true optimal solution. LP-based approximation algorithms exploit such reflected structures to find good solutions to hard optimization problems. In this section, we will survey how two main LP-based methodologies, LP-rounding and the primal-dual method, have been applied to facility location and clustering problems.

As LP-based algorithms rely on the fact that the LP optimum is a lower bound (for minimization problems) on the optimum, it is important that the LP relaxation is strong enough to provide a close bound. Given an LP relaxation, we therefore investigate its *integrality gap*, defined as the worst-case (supremum) ratio between the true optimum and fractional LP optimum. Naturally (but not necessarily), many LP-based algorithms provide an upper bound on the integrality gap, and we are often also interested in the lower bounds as they suggest a "limit" on the LP-based methods relying on that particular relaxation. Following is the standard LP relaxation with a constant integrality gap, used by many LP-based algorithms for the facility location problem.

$$
\begin{aligned}
\text{minimize } & \sum_{i \in \mathcal{F}} f_i y_i + \sum_{i \in \mathcal{F}, j \in \mathcal{D}} c(i, j) x_{ij}, \\
\text{subject to } & \sum_{i \in \mathcal{F}} x_{ij} \geq 1, & \forall j \in \mathcal{D}, \\
& x_{ij} \leq y_i, & \forall i \in \mathcal{F}, j \in \mathcal{D}, \\
& \mathbf{x}, \mathbf{y} \geq \mathbf{0}.
\end{aligned}
$$

We call x_{ij}'s *connection variables* as they encode where each client is assigned; y_i's are called *opening variables* as they encode which facilities are to be opened.

3.1 LP-Rounding for Facility Location

The LP-rounding method is an approach where we first solve an LP relaxation of the given problem, and then "round" the optimal solution into an integral solution while ensuring that the cost does not increase too much.

The first $O(1)$-approximation algorithm for the facility location problem was given by Shmoys, Tardos, and Aardal [39] via LP-rounding. Their algorithm first solves the standard LP relaxation, and then apply a technique called *filtering* on the connection variables: for each client, we consider all its connection variables and compute the average connection cost, weighted by the connection variables. Then we discard all connection variables whose assignment distance is too large compared to this average. From Markov's inequality, we know that the sum of the remaining connection variables is bounded from below by a constant (as opposed to 1 before the filtering), and therefore by scaling these variables up, we can restore LP feasibilty by losing a constant factor in the cost. In exchange, we obtain a useful property that, for each client, the assignment distances of nonzero connection variables are all approximately equal.

Once we are finished with this filtering process, we greedily construct a clustering of facilities: we consider every client in the increasing order of the assignment distance. For each client j, if none of the facilities to which j is fractionally assigned has been clustered, we take j and all these facilities as a new cluster. Note that some facilities may not belong to any cluster. For each cluster, we choose a facility with the minimum opening cost and open it. As was observed earlier, once the set of open facilities is specified, the optimal assignment can be computed and hence this completes the algorithm description. (The analysis, however, also provides a good assignment that can be used without having to compute the optimal assignment.)

Theorem 8 (Shmoys et al. [39]) *There exists a 3.16-approximation algorithm for the facility location problem.*

Proof (sketch) Recall that the filtering process loses a constant factor in the approximation ratio but maintains the LP feasibility. First we argue that, towards the opening cost, we do not pay more than what the LP pays. For each cluster centered at $j \in \mathcal{D}$, we open a min-cost facility, and these are all the facilities that are opened. The LP constraints require that the sum of the connection variables of j is at least 1, and therefore LP pays at least as much as the algorithm does.

We now focus on the connection cost. For each cluster center j, we do not pay much more than what the LP pays since j can be assigned to the min-cost facility of this cluster; recall that the assignment distances are approximately equal around each client. For a client j that is *not* a cluster center, the fact that j could not become a cluster center implies that there is another client center j' whose connection cost is smaller, and j' has claimed a facility i to which j is also fractionally assigned. We will assign j to the same facility j' is assigned to, say i', and we bound this assignment distance by a length-3 path $j - i - j' - i'$. Note that $c(i, j)$ is already paid for by the LP. Moreover, since the connection cost of j' is smaller, the $c(i, j')$ and $c(i', j')$ can also be charged against $c(i, j)$. Thus, the algorithm does not use more than approximately three times what the LP has paid.

These arguments, with a more careful analysis of the constants, show that the given algorithm is a 3.16-approximation algorithm for the facility location problem. □

Note that the LP may pay more opening cost than what the algorithm pays, since the algorithm blindly chooses the min-cost facility in each cluster. Chudak and Shmoys [17] interpreted the connection variables of each cluster center as probabilities and randomly chose the facility to open, giving a $(1 + \frac{2}{e})$-approximation algorithm. Further improvements due to Byrka [9] gives a 1.6774-approximation algorithm, and since this algorithm approximates the connection cost much better than by the factor of 1.6774, it can be combined with primal-dual algorithms to give an even better performance guarantee.

3.2 Primal-Dual Algorithms for Facility Location

The primal-dual method is an approach where an LP relaxation is not explicitly solved. Instead, the algorithm usually constructs a primal feasible solution in a purely combinatorial way, while constructing its near-optimality certificate in the form of a feasible dual solution. Note that a *feasible* dual solution provides a lower bound on the optimum. This process is usually guided by the complementary slackness: we often start with a trivial dual feasible solution (e.g. all zeroes) along with a primal infeasible solution. We then increase some of the dual variables until a dual constraint becomes tight. At this point, the complementary slackness indicates which primal variable is a good candidate to increase. This process is repeated until the primal feasibility is attained. The tightness of dual constraints often implies that the increase in the primal and dual cost are within a constant factor of each other, giving a performance guarantee. Unsurprisingly, the perfect complementary slackness is not usually acheived for NP-hard problems, as that would imply that the LP is integral.

Jain and Vazirani [28] gave the first primal-dual approximation algorithm for the facility location problem. In order to facilitate describing this algorithm, we show the dual of the standard LP relaxation below.

$$\text{maximize } \sum_{j\in\mathcal{D}} \alpha_j,$$
$$\text{subject to } \alpha_j - \beta_{ij} \le c(i, j), \ \forall i \in \mathcal{F}, j \in \mathcal{D},$$
$$\sum_{j\in\mathcal{D}} \beta_{ij} \le f_i, \quad \forall i \in \mathcal{F},$$
$$\alpha, \beta \ge \mathbf{0}.$$

The algorithm starts with the empty primal solution and the zero dual solution. We uniformly increase the dual variables of all unassigned clients as the time passes, and once a client is assigned, its dual stops increasing. Thus, at any given time t, the dual variable of an unassigned client will have the value of t. Initially, all clients are unassigned.

Once α_j reaches $c(i, j)$ for some (i, j), we say that the edge is *tight* and start increasing β_{ij} at the same rate so that $\alpha_j - \beta_{ij}$ remains equal to $c(i, j)$, maintaining the dual feasibility. If a tight edge (i, j) further satisfies $\beta_{ij} > 0$, we say that the edge is *special*. The other dual constraints that can become tight as the algorithm progresses are the second set of constraints. When $\sum_{j\in\mathcal{D}} \beta_{ij}$ reaches f_i, we declare i *temporarily opened*, and assign to i every client j such that (i, j) is tight. Later in the algorithm, if (i, j') becomes tight for some client j', j' will be immediately assigned to i. The process ends when all clients become assigned.

The algorithm then enters the second phase: we consider a graph on the temporarily opened facilities, in which two facilities i and i' are adjacent if there exists a client

j such that (i, j) and (i', j) are both special. We compute a maximal independent set of this graph, and this is the final set of facilities to be opened.

Theorem 9 (Jain and Vazirani [28]) *There exists a 3-approximation algorithm for the facility location problem.*

Proof (sketch) It is easy to verify that the algorithm runs in polynomial time. We will therefore focus on bounding the performance guarantee of this algorithm.

A facility becomes temporarily opened only when the corresponding dual constraint becomes tight; hence, in order to bound the opening cost paid by the algorithm, it suffices to ensure that each client contributes towards at most one facility opening. (Note that there can be multiple number of temporarily opened facilities to which a given client contributes.) We ensure this by taking an independent set during the second phase of the algorithm, since all the facilities to which any given client makes positive contribution are mutually adjacent in the constructed graph.

Now it remains to bound the connection cost. For each client $j \in \mathcal{D}$, if there exists an open facility $i \in \mathcal{F}$ such that (i, j) is tight, then the dual variable α_j is sufficient to pay the connection cost of j. If there exists no such open facility, then since j was assigned during the first phase of the algorithm, there is some facilty i such that (i, j) is tight and i was temporarily opened but it was not chosen during the second phase. Note that $c(i, j)$ again can be paid by the dual variable. From the maximality of the independent set, there must have been another facility i' that *is* chosen in the second phase and adjacent to i in the constructed graph. From the construction, there exists some client j' such that both (i, j') and (i', j') are special; we therefore have that i and i' are within the distance of $2\alpha_{j'}$. Finally, we can show that j' was assigned at a no later time than j because (i', j') is tight. This shows that the connection cost of every client j is bounded by $3\alpha_j$ in all cases. □

Jain et al. [27] give an improved primal-dual algorithm for the facility location problem.

Theorem 10 (Jain et al. [27]) *There exists a 1.61-approximation algorithm for the facility location problem.*

Finally, combining LP-rounding algorithms with the primal-dual method, Li [34] gives the best approximation ratio known.

Theorem 11 (Li [34]) *There exists a 1.488-approximation algorithm for the facility location problem.*

3.3 *k-median and Its Relaxations*

The first constant approximation algorithm for k-median is due to Charikar et al. [15]. Their $\frac{20}{3}$-approximation algorithm is based on a rounding of the standard LP, which now needs to be modified as follows:

$$\text{minimize } \sum_{i \in \mathcal{F}, j \in \mathcal{D}} c(i, j) x_{ij},$$
$$\text{subject to } \sum_{i \in \mathcal{F}} x_{ij} \geq 1, \qquad \forall j \in \mathcal{D},$$
$$x_{ij} \leq y_i, \qquad \forall i \in \mathcal{F}, j \in \mathcal{D},$$
$$\sum_{i \in \mathcal{F}} y_i \leq k,$$
$$\mathbf{x}, \mathbf{y} \geq \mathbf{0}.$$

Theorem 12 (Charikar et al. [15]) *There exists a $\frac{20}{3}$-approximation algorithm for k-median based on rounding the standard LP.*

Several of the ideas in [15] are inspired from the $O(1)$-approximation algorithms for the uncapacitated facility location problem that we discussed in previous subsections. By further developing these techniques, Charikar and Li [16] obtained an improved algorithm based on LP-rounding:

Theorem 13 (Charikar and Li [16]) *There exists a 3.25-approximation algorithm for k-median based on rounding the standard LP.*

Even though the above results use techniques that are close to those used for facility location, the current best algorithms for k-median use an even more direct connection. As was pointed out earlier, the facility location problem (with uniform opening costs) is a Lagrangian relaxation of the k-median problem. This connection can also be motivated by basic economic theory: if we let the opening costs of facilities be small then a "good" solution to uncapacitated facility location will open many facilities whereas if we let the opening costs of facilities be large then a good solution will only open few facilities. By appropriately selecting the cost of facilities, one can therefore expect that an algorithm for uncapacitated facility location opens close to k facilities and therefore almost gives a solution to the k-median problem. A key difficulty here is that this approach will only give a solution that approximately opens k facilities. Nevertheless, Jain and Vazirani [28] first exploited this concept in their beautiful paper and showed how to overcome this issue by further losing a factor 2 in the approximation guarantee. They then combined this with their 3-approximation algorithm for uncapacited facility location that we described in Sect. 3.2, to obtain a 6-approximation algorithm for k-median.

Theorem 14 (Jain and Vazirani [28]) *There exists a 6-approximation algorithm for the k-median problem.*

The factor 3 was later improved by Jain et al. [27] to 2, resulting in a 4-approximation algorithm for k-median based on the primal-dual method:

Theorem 15 (Jain et al. [27]) *There exists a 4-approximation algorithm for the k-median problem.*

We remark that we cannot use any arbitrary approximation algorithm for facility location in conjunction with this framework. We say a ρ-approximation algorithm for the facility location problem is *Lagrangian multiplier preserving* if the cost of the algorithm's output is at most ρ times the optimum, even when we redefine the cost of the algorithm's output so that the algorithm pays ρ times the normal opening cost.

The framework of Jain and Vazirani [28] requires a Lagrangian multiplier preserving algorithm; this is why we could not use the 1.488-approximation algorithm due to Li [34] in place of the 2-approximation algorithm of Jain et al. [27].

At a first sight, it seems hard to get any significant improvements using the above framework: it is NP-hard to achieve an approximation ratio that is better than $(1 + 2/e)$ by a Lagrangian multiplier preserving approximation algorithm, and the factor 2 that we lose in order to convert such an algorithm into an approximation algorithm for the k-median problem is known to be tight. Hence, the best possible guarantee conceivable within this framework is $2 \cdot (1 + 2/e) \approx 3.47$, which is worse than the guarantees given by the local search method (Theorem 5) and LP-rounding (Theorem 13). The key result of Li and Svensson [36] to overcome this issue is the following:

Theorem 16 (Li and Svensson [36]) *Let $\mathcal{A}$ be an α-approximation algorithm for k-median that opens at most $k + c$ facilities. Then, for every $\varepsilon > 0$, there exists an $(\alpha + \varepsilon)$-approximation algorithm for k-median that opens at most k facilities whose running time is $n^{O(c/\varepsilon)}$ times the running time of $\mathcal{A}$.*

Informally, their theorem says that an algorithm that approximately opens k facilities can be turned into an algorithm that opens exactly k facilities without deteriorating the approximation guarantee. This theorem frees us from the barrier of 3.47 discussed above, since allowing the algorithm to open just slightly more than k facilities (even $k + 1$ suffices) invalidates the tightness of the factor of 2 that we lose in the framework of Jain and Vazirani [28]. In fact, Li and Svensson [36] showed that if we open a few more facilities then the factor 2 can be improved to $\frac{1+\sqrt{3}}{2}$. Combining this with the above theorem and the 2-approximation algorithm for facility location by Jain et al. [27] yields a $(1 + \sqrt{3} + \varepsilon)$-approximation algorithm for k-median for any $\varepsilon > 0$, improving upon the previous best known ratio of 3 given by local search. Using a careful dependent rounding, the factor $\frac{1+\sqrt{3}}{2}$ was subsequently improved by Byrka et al. [12] to obtain the best approximation guarantee known for k-median:

Theorem 17 (Byrka et al. [12][6]) *For every $\varepsilon > 0$, there is a $(2.675 + \varepsilon)$-approximation algorithm for k-median.*

3.4 Capacitated Problems

While the LP-based methods grant us a nearly tight understanding of the approximability of the facility location problem, they have not been as successful in capacitated problems. In fact, for the capacitated facility location problem and the capacitated k-center problem, the first LP-based $O(1)$-approximation algorithms were

---

[6]We remark that the conference version of this paper claimed a guarantee of $2.611 + \varepsilon$ where the additional improvement came from an improved Lagrangian multiplier preserving algorithm for facility location. The up-to-date arXiv version of the same paper [11] however noted that this part of their improvement unfortunately contained an error, changing the final approximation guarantee to $2.675 + \varepsilon$.

only recently obtained. For the capacitated k-median problem, we do not know of an $O(1)$-approximation algorithm at all.

In the capacitated facility location problem, the difficulty stems from the fact that the standard LP relaxation, augmented with the natural capacity constraints

$$\sum_{j\in\mathcal{D}} x_{ij} \le U_i y_i, \quad \forall i \in \mathcal{F},$$

has an unbounded integrality gap. In fact, the example showing this fact is so simple that we can assume c to be a constant zero function. For an arbitrary integer N, consider an instance with $N + 1$ clients and two facilities, all of which are "at the same point", i.e., $c(i, j)$ is zero for all i and j. Both facilities have the capacity of N, whereas the opening cost of one facility is 0 and the other is 1. It is easy to verify that the true optimum is of cost 1, whereas the LP optimum is of value $1/N$ for this instance.

The first LP-based $O(1)$-approximation algorithm for the capacitated facility location problem was obtained as recent as in 2014 [5]. This result inevitably accompanied a new LP relaxation, where the new LP relaxation was of similar flavor as *knapsack-cover inequalities* [13, 42]. The new relaxation considers a multicommodity flow network where each client generates a unit flow of its own commodity type, and each facility becomes a sink of capacity U_i where this capacity is to be shared by all the commodities and capacity 1 where this capacity is to be used by each individual commodity. We introduce arcs from clients to facilities whose capacities are given by the connection variables. The constraint that says this multicommodity flow network should be feasible corresponds to the standard LP relaxation, and An et al. [5] strengthen this relaxation by considering every possible "partial assignment" of clients to facilities and introducing *backward arcs* from facilities to clients corresponding to this partial assignment. An et al. show that the constraint that this multicommodity flow network with adjusted sink capacities and commodity demands should also be feasible is a valid constraint, since the backward arcs allow the partial assignment to be "undone" if necessary in order to restore feasibility.

Based on this relaxation, they reduce the capacitated facility location problem to the *soft-capacitated facility location problem*, in which the capacity constraints are allowed to be violated by a provable factor. Abrams et al. [1] give an $O(1)$-approximation algorithm for the soft-capacitated facility location problem based on the standard LP relaxation, leading to the following result.

Theorem 18 (An et al. [5]) *There exists an LP-based $O(1)$-approximation algorithm for the capacitated facility location problem.*

The uncapacitated k-center problem has a simple greedy 2-approximation algorithm that is the best possible [25, 26]. However, the problem becomes difficult when it is capacitated: the first $O(1)$-approximation algorithm for the capacitated k-center problem was given only in 2012 [20]. As the k-center problem is a bottleneck optimization problem, it is natural to attempt to solve this problem by using the standard technique: we guess the optimum τ, and construct an unweighted graph where edges exist only between a facility and a client whose assignment distance is at most τ. We then use the standard LP relaxation to check if our guess is correct: we treat the standard LP as a feasibility LP, and allow only the connection variables whose corresponding edges exist in the unweighted graph to have nonzero values. If the LP is infeasible, we have certified that our guess was incorrect; otherwise we give an algorithm that transforms the fractional LP solution into an integral solution where an assignment occurs only between facilities and clients that are close to each other in the unweighted graph. It is easy to show that, if we can find such an algorithm, then the framework provides an approximation algorithm; however, there exist examples that show the LP is too weak: for some graphs, the LP is feasible but there does not exist any integral solution whose assignment is only between close nodes.

Cygan et al. [20] observed that it suffices to provide such an algorithm only for connected graphs, and showed that the LP is strong enough if we restrict ourselves to connected graphs. Based on this observation, they give the first $O(1)$-approximation algorithm.

Theorem 19 (Cygan et al. [20]) *There exists an $O(1)$-approximation algorithm for the capacitated k-center problem.*

An et al. [4] give a simplified algorithm with a better performance guarantee.

Theorem 20 (An et al. [4]) *There exists a 9-approximation algorithm for the capacitated k-center problem.*

For the capacitated k-median problem, in contrast, we do not know of any $O(1)$-approximation algorithm. However, there has been recent progress when we relax the problem so that the algorithm is allowed to violate the capacities or the hard bound k on the number of open facilities. Improving upon work by Byrka et al. [10], Demirci and Li [21] gave an algorithm that slightly violates the capacities:

Theorem 21 (Demirci and Li [21]) *For all $\varepsilon > 0$, there exists a polynomial-time algorithm that returns a solution to the capacitated k-median problem which violates the capacities by a factor of $1 + \varepsilon$, where the cost of the returned solution is no worse than $O(1/\varepsilon^5)$ times the true (unrelaxed) optimum.*

Li [35], on the other hand, gave an algorithm that violates the number of opened facilities instead; the algorithm does not violate the capacities but may open the same facility multiple times.

Theorem 22 (Li [35]) *For all $\varepsilon > 0$, there exists a polynomial-time algorithm that returns a solution to the capacitated k-median problem which violates the number of open facilities by a factor of $1 + \varepsilon$ and the same facility may be opened multiple times, where the cost of the returned solution is no worse than $O(\frac{1}{\varepsilon^2} \log \frac{1}{\varepsilon})$ times the true (unrelaxed) optimum.*

Table 1 Approximability bounds for clustering problems

Problem	Approximation ratio	Hardness of approximation
Facility location	1.488 [34]	1.463 [24]
k-median	2.675 [12]	$1 + 2/e$ [27]
k-means [a]	$6.357 + \varepsilon$ [3]	1.0013 [33]
k-center	2 [25]	$2 - \varepsilon$ [26]

[a] The continuous, Euclidean problem considered

4 Future Directions

One of the most fundamental open questions in this field is to determine the exact approximability of facility location, k-median, and k-means. In Table 1, we summarize the best approximation ratios and hardness results known for these problems.

The largest gap in our understanding is for k-means. However, even for the facility location problem it appears that fundamentally new techniques will be needed to close the gap: the 1.488-approximation algorithm was obtained by combining previous approaches and "squeezing the best" via randomized choice of parameters and it seems unlikely to get further improvements in a similar vein.

Even though it is conceivable that the current algorithms are indeed the best possible, we tend to believe that this is not the case and new algorithmic ideas are needed to settle the approximability of these problems. We will conclude this survey with some concrete open problems in this line of research.

Open Problem 1. *Design non-oblivious local search algorithms for the mentioned problems with improved guarantees.*

A non-oblivious local search is a local search that does not simply use the objective function of the problem to define the improving steps. The local search algorithms mentioned in Sect. 2 are often the best possible for oblivious local search algorithms, but no such bounds are known for non-oblivious local search algorithms.

Open Problem 2. *Design a Lagrangian multiplier preserving $(1 + 2/e)$-approximation algorithm for the facility location problem.*

In addition to closing the gap between the approximability bounds for the facility location problem, solving this problem may lead to an improved approximation algorithm for the k-median problem as well. We remark that the best algorithm known for k-median that opens k facilities in expectation achieves a factor of 2, due to Jain et al. [27].

Open Problem 3. *Design an LP-rounding algorithm for k-median with an improved guarantee, where the algorithm is allowed to open $k + c$ facilities for some constant c.*

Here, we are interested in exploiting Theorem 16 by using a more direct LP-rounding instead of using the connection to facility location. We remark that the worst integrality gap of the standard LP known for k-median, if we are allowed to open $k + 1$ facilities, is $1 + 2/e$ which matches the best hardness result known.

Open Problem 4. *Narrow the gap in our understanding of k-means.*

The difficulty of this problem is that the distances are squared and thus we cannot anymore rely on the triangle inequality without losing a large factor in the approximation guarantee. At the same time, no strong hardness results are known for the k-means problem. Some of the difficulties for proving hardness results are that it is a continuous problem and the distances form an Euclidean metric. Indeed, if we consider the more general version where we have a discrete metric and a finite set of facilities, then it is easy to see that the same reduction as that for k-median gives a hardness factor of $1 + 8/e$ for this general version of k-means. We believe that new techniques are needed to gain a better understanding of the k-means problem.

Capacitated variants only recently started to see improvements, and devising algorithms that better approximate these problems is another interesting future direction. Improving the algorithms for the capacitated facility location problem and the capacitated k-center problem is an interesting open question; but in particular, the k-median problem stands out among these capacitated problems, as no $O(1)$-approximation algorithm is known to this date. Interestingly, in spite of the difficulty of designing good approximation algorithms, we have no stronger hardness results in the presence of capacities. Apart from obvious open questions, following recent results, we present some concrete open problems below.

Open Problem 5. *Settle the integrality gap of the relaxation proposed in An et al.* [5] *for the capacitated facility location problem.*

It is consistent with our current knowledge that the proposed relaxation has an integrality gap of 2; proving this would most likely lead to an improved algorithm for capacitated facility location.

Open Problem 6. *Does the capacitated facility location problem have a compact (polynomial-sized) relaxation with a constant integrality gap?*

The current relaxation is of exponential size and is approximately solved using a relaxed separation oracle. A potentially easier problem that is still interesting is to devise a relaxation of the capacitated facility location problem with a constant integrality gap that can be solved in polynomial time.

Finally, here is an open problem that is less well-defined: describe a systematic way for strengthening the standard LP relaxation for capacitated k-center and capacitated facility location so that the resulting relaxation has a constant integrality gap. While this question may appear vague, we feel that the current relaxation for the capacitated facility location problem is rather specialized to that problem; one can also obtain specialized relaxations for the capacitated k-center problem with constant integrality gaps. However, we believe that a more systematic approach that can be shared by these multiple problems would give new insights. The ultimate hope would be that the new inequalities might help obtaining the first $O(1)$-approximation algorithm for the capacitated k-median problem, which remains a main open question for capacitated problems.

Acknowledgements We thank the anonymous reviewer of this article for the helpful comments. This work was supported by the National Research Foundation of Korea (NRF) grant funded by the Korea government (MSIP) (No. 2016R1C1B1012910), a Yonsei University New Faculty Seed Grant, and ERC Starting Grant 335288-OptApprox.

References

1. Z. Abrams, A. Meyerson, K. Munagala, S. Plotkin, On the integrality gap of capacitated facility location. Technical Report CMU-CS-02-199 (Carnegie Mellon University, 2002), http://theory. stanford.edu/~za/CapFL/CapFL.pdf
2. A. Aggarwal, A. Louis, M. Bansal, N. Garg, N. Gupta, S. Gupta, S. Jain, A 3-approximation algorithm for the facility location problem with uniform capacities. Math. Progr. **141**(1–2), 527–547 (2013)
3. S. Ahmadian, A. Norouzi-Fard, O. Svensson, J. Ward, Better guarantees for k-means and euclidean k-median by primal-dual algorithms (2016), http://arxiv.org/abs/1612.07925. CoRR abs/1612.07925
4. H.C. An, A. Bhaskara, C. Chekuri, S. Gupta, V. Madan, O. Svensson, Centrality of trees for capacitated k-center. Math. Progr. **154**(1), 29–53 (2015). doi:10.1007/s10107-014-0857-y
5. H.C, An, M. Singh, O. Svensson, LP-based algorithms for capacitated facility location, in *Proceedings of the 55th Annual Symposium on Foundations of Computer Science (FOCS)* (2014), pp. 256–265
6. V. Arya, N. Garg, R. Khandekar, A. Meyerson, K. Munagala, V. Pandit, Local search heuristic for k-median and facility location problems, in *Proceedings of the Thirty-third Annual ACM Symposium on Theory of Computing, STOC 2001*, pp. 21–29. doi:10.1145/380752.380755
7. M.L. Balinski, On finding integer solutions to linear programs, in *Proceedings of the IBM Scientific Computing Symposium on Combinatorial Problems* (1966), pp. 225–248
8. M. Bansal, N. Garg, N. Gupta, A 5-approximation for capacitated facility location, in *Proceedings of the 20th European Symposium on Algorithms (ESA)* (2012), pp. 133–144
9. J. Byrka, An optimal bifactor approximation algorithm for the metric uncapacitated facility location problem, in *APPROX-RANDOM* (2007), pp. 29–43
10. J. Byrka, K. Fleszar, B. Rybicki, J. Spoerhase, Bi-factor approximation algorithms for hard capacitated k-median problems, in *Proceedings of the Twenty-Sixth Annual ACM-SIAM Symposium on Discrete Algorithms, SODA 2015*, pp. 722–736, http://dl.acm.org/citation.cfm?id= 2722129.2722178
11. J. Byrka, T. Pensyl, B. Rybicki, A. Srinivasan, K. Trinh, An improved approximation for k-median, and positive correlation in budgeted optimization (2014), http://arxiv.org/abs/1406. 2951. CoRR abs/1406.2951
12. J. Byrka, T. Pensyl, B. Rybicki, A. Srinivasan, K. Trinh, An improved approximation for k-median, and positive correlation in budgeted optimization, in *Proceedings of the Twenty-Sixth Annual ACM-SIAM Symposium on Discrete Algorithms, SODA 2015*, pp. 737–756, http://dl. acm.org/citation.cfm?id=2722129.2722179

13. R.D. Carr, L.K. Fleischer, V.J. Leung, C.A. Phillips, Strengthening integrality gaps for capacitated network design and covering problems, in *Proceedings of the 11th Annual ACM-SIAM Symposium on Discrete Algorithms (SODA)* (2000), pp. 106–115

14. M. Charikar, S. Guha, Improved combinatorial algorithms for facility location problems. SIAM J. Comput. **34**(4), 803–824 (2005)

15. M. Charikar, S. Guha, E. Tardos, D.B. Shmoys, A constant-factor approximation algorithm for the k-median problem (extended abstract), in *Proceedings of the Thirty-first Annual ACM Symposium on Theory of Computing, STOC 1999*, pp. 1–10. doi:10.1145/301250.301257

16. M. Charikar, S. Li, *A Dependent LP-Rounding Approach for the k-Median Problem* in *39th International Colloquium on Automata, Languages, and Programming, ICALP 2012*, pp. 194–205. doi:10.1007/978-3-642-31594-7_17

17. F.A. Chudak, D.B. Shmoys, Improved approximation algorithms for the uncapacitated facility location problem. SIAM J. Comput. **33**(1), 1–25 (2004)

18. F.A. Chudak, D.P. Williamson, Improved approximation algorithms for capacitated facility location problems. Math. Progr. **102**(2), 207–222 (2005)

19. V. Cohen-Addad, P.N. Klein, C. Mathieu, Local search yields approximation schemes for k-means and k-median in euclidean and minor-free metrics, in *Proceedings of the 57nd Annual Symposium on Foundations of Computer Science (FOCS)* (2016), pp. 353–364

20. M. Cygan, M. Hajiaghayi, S. Khuller, LP rounding for k-centers with non-uniform hard capacities, in *FOCS* (2012), pp. 273–282

21. H.G. Demirci, S. Li, Constant approximation for capacitated k-median with (1+eps.)-capacity violation, in *43rd International Colloquium on Automata, Languages, and Programming, ICALP 2016* (2016), pp. 73:1–73:14

22. G. Diehr, An algorithm for the p-median problem. Technical Report No. 191 (Western Management Science Institute, UCLA 1972)

23. Z. Friggstad, M. Rezapour, M. Salavatipour, Local search yields a PTAS for k-means in doubling metrics, in *Proceedings of the 57nd Annual Symposium on Foundations of Computer Science (FOCS)* (2016), pp. 365-374

24. S. Guha, S. Khuller, Greedy strikes back: Improved facility location algorithms. J. Algorithms **31**(1), 228–248 (1999)

25. D.S. Hochbaum, D.B. Shmoys, A best possible heuristic for the k-center problem. Math. Oper. Res. **10**, 180–184 (1985)

26. W.L. Hsu, G.L. Nemhauser, Easy and hard bottleneck location problems. Discret. Appl. Math. **1**(3), 209–215 (1979). doi:10.1016/0166-218X(79)90044-1, http://www.sciencedirect.com/science/article/pii/0166218X79900441

27. K. Jain, M. Mahdian, A. Saberi, A new greedy approach for facility location problems, in *Proceedings of the 34th annual ACM symposium on Theory of computing (STOC)* (2002), pp. 731–740

28. K. Jain, V.V. Vazirani, Approximation algorithms for metric facility location and k-median problems using the primal-dual schema and Lagrangian relaxation. J. ACM **48**(2), 274–296 (2001)

29. T. Kanungo, D.M. Mount, N.S. Netanyahu, C.D. Piatko, R. Silverman, A.Y. Wu, An efficient k-means clustering algorithm: analysis and implementation. IEEE Trans. Pattern Anal. Mach. Intell. **24**(7), 881–892 (2002). doi:10.1109/TPAMI.2002.1017616

30. M.R. Korupolu, C.G. Plaxton, R. Rajaraman, Analysis of a local search heuristic for facility location problems, in *Proceedings of the Ninth Annual ACM-SIAM Symposium on Discrete Algorithms, SODA 1998*, pp. 1–10, http://dl.acm.org/citation.cfm?id=314613.314616

31. M.R. Korupolu, C.G. Plaxton, R. Rajaraman, Analysis of a local search heuristic for facility location problems. J. Algorithms **37**(1), 146–188 (2000)

32. A.A. Kuehn, M.J. Hamburger, A heuristic program for locating warehouses. Manag. Sci. **9**(4), 643–666 (1963)

33. E. Lee, M. Schmidt, J. Wright, Improved and simplified inapproximability for k-means (2015), arXiv:1509.00916

34. S. Li, A 1.488 approximation algorithm for the uncapacitated facility location problem. Inf. Comput. **222**(0), 45–58 (2013). doi:10.1016/j.ic.2012.01.007, http://www.sciencedirect.com/science/article/pii/S0890540112001459
35. S. Li, Approximating capacitated k-median with $(1 + \varepsilon)k$ open facilities, in *Proceedings of the Twenty-Seventh Annual ACM-SIAM Symposium on Discrete Algorithms, SODA 2016*, pp. 786–796, http://dl.acm.org/citation.cfm?id=2884435.2884491
36. S. Li, O. Svensson, Approximating k-median via pseudo-approximation, in *Proceedings of the Forty-fifth Annual ACM Symposium on Theory of Computing, STOC 2013*, pp. 901–910. doi:10.1145/2488608.2488723
37. A.S. Manne, Plant location under economies-of-scale-decentralization and computation. Manage. Sci. **11**(2), 213–235 (1964). doi:10.1287/mnsc.11.2.213
38. M. Pál, É. Tardos, T. Wexler, Facility location with nonuniform hard capacities, in *Proceedings of the 42nd Annual Symposium on Foundations of Computer Science (FOCS)* (2001), pp. 329–338
39. D.B. Shmoys, E. Tardos, K. Aardal, Approximation algorithms for facility location problems (extended abstract), in *Proceedings of the 29th annual ACM symposium on Theory of computing (STOC)* (1997), pp. 265–274
40. J.F. Stollsteimer, A working model for plant numbers and locations. J. Farm Econ. **45**(3), 631–645 (1963), http://www.jstor.org/stable/1235442
41. M.B. Teitz, P. Bart, Heuristic methods for estimating the generalized vertex median of a weighted graph. Oper. Res. **16**(5), 955–961 (1968). doi:10.1287/opre.16.5.955
42. L.A. Wolsey, Faces for a linear inequality in 0–1 variables. Math. Progr. **8**(1), 165–178 (1975). doi:10.1007/BF01580441

Graph Stabilization: A Survey

Karthekeyan Chandrasekaran

Abstract Graph stabilization has raised a family of network design problems that has received considerable attention recently. Stable graphs are those graphs for which the matching game has non-empty core. In the optimization terminology, they are graphs for which the fractional matching linear program has an integral optimum solution. Graph stabilization involves minimally modifying a given graph to make it stable. In this survey, we outline recent developments in graph stabilization and highlight some open problems.

1 Introduction

The increasingly networked structure of human interactions in modern society has raised fascinating and novel game-theoretic questions [11, 37, 41]. Graph models for such networked interactions have been a central topic of research in algorithmic game theory over the last two decades [15, 28]. Fundamental to these interactions is matching, namely players in a network pairing up according to some self-interest. Formally, a matching in a graph is a collection of edges such that each vertex belongs to at most one edge in the collection. Owing to the simplicity of the description and the widespread nature of combinatorial objects that they can model, matchings have attracted much interest in games over networks [6, 8, 9, 18–20, 24, 26, 31, 40]. They play a prominent role in the network bargaining game introduced by Kleinberg and Tardos [31] as a network generalization of Nash's two-player bargaining game [36] as well as the classic cooperative matching game introduced by Shapley and Shubik [40].

An instance of a *cooperative matching game* is specified by an edge-weighted graph. The vertices correspond to players. The weight of an edge between two players represents the profit achievable from the relationship. The value of a coalition of players is the weight of the maximum matching achievable in the subgraph induced by the players in the coalition. The profit of the game is the value of the grand-

K. Chandrasekaran (✉)
University of Illinois, Urbana-Champaign, IL, USA
e-mail: karthe@illinois.edu

© Springer Nature Singapore Pte Ltd. 2017
T. Fukunaga and K. Kawarabayashi (eds.), *Combinatorial Optimization and Graph Algorithms*, DOI 10.1007/978-981-10-6147-9_2

coalition, namely the maximum weight of a matching in the graph. The central network authority needs to distribute the profit of the game among the players so that the cumulative allocation to any coalition is at least its value. Such profit-sharing allocations are known as *core allocations* [23]. The existence of a core allocation ensures that no coalition can achieve larger profit than their allocated value by acting for themselves and hence, the stability of the grand coalition. Core allocations are desirable for the fundamental reason that they foster cooperation among the players.

Core allocations do not exist for every graph. For instance, consider a cycle on three vertices with unit-weight on the edges: any distribution of the value of the maximum matching in the graph, namely 1, among the three players leaves a pair of players who are cumulatively receiving less than 1, which is the value of the coalition formed by that pair. Graphs which have a core allocation are known as *stable graphs*. Since stable graphs are the only graphs for which the grand coalition of players cooperate, they are particularly attractive from the perspective of the central network authority. For this reason, when faced with an unstable graph, the central network authority is interested in stabilizing it, i.e., modifying it in the *least intrusive fashion* in order to convert it into a stable graph.

In this survey, we will describe recent progress in graph stabilization. The main goals of the survey are to provide game-theoretic interpretations of stabilization models and intermediate structural results, highlight the interplay between graph-theory and mathematical programming as algorithmic techniques towards stabilization, and state some immediate open problems to spur further research. The results to be presented will include stabilization by edge and vertex deletion, edge and vertex addition and edge weight addition [2, 10, 12, 27] (see Table 1 for a summary).

We assume familiarity with integer programming (IP), linear programming (LP), LP-duality (see Schrijver [38]), approximation terminology (see Vazirani [44], Shmoys-Williamson [46]), Edmonds' maximum matching algorithm [13] and basic graph theory (see West [45]). For an elaborate theory of matchings, we refer the reader to the textbooks by Lovász-Plummer [34] and Schrijver [39].

1.1 Definitions

Throughout, we are only interested in undirected graphs. We will denote an edge-weighted graph $G = (V, E)$ with edge weights $w : E \to \mathbb{R}_+$ by (G, w). For a vertex $u \in V$, let $\delta(u)$ denote the set of edges incident to u. For a subset $S \subseteq V$, let $E[S]$ denote the set of edges both of whose end vertices are in S, let $G[S]$ denote the subgraph induced by the vertices in S and $N_G(S)$ denote the set of vertices in $V \setminus S$ adjacent to at least one vertex in S. We denote the weighted subgraph induced by a subset $S \subseteq V$ by $(G[S], w)$. For given costs on edges (vertices), the cost of a subset S of edges (vertices) is the sum of the costs on the edges (vertices) in S. A matching in G is a subset M of edges such that each vertex is incident to at most one edge in M. The problem of finding a matching of maximum weight in (G, w) is formulated by the following integer program:

Table 1 Stabilization problems for unit-weight graphs. See Sects. 3, 4 and 5 for model descriptions

Model	Hardness	Approximation
min-EDGE-DEL	$(2 - \varepsilon)$-inapprox for $\varepsilon > 0$	$O(\omega)$-approx in ω-sparse graphs, 2-approx in regular graphs
min-EDGE-ADD	P	–
min-Cost-EDGE-ADD	NP-hard	
max-EDGE-SUBGRAPH	NP-hard	$(5/3)$-approx
min-VERTEX-DEL	P	–
min-VERTEX-ADD	P	–
min-Cost-VERTEX-DEL	NP-hard	$(|C| + 1)$-approx
max-Cost-VERTEX-SUBGRAPH	NP-hard	2-approx
min-EDGE-WT-ADD	$c \log |V|$-inapprox for some constant c, $O(|V|^{\frac{1}{16} - \varepsilon})$-inapprox for $\varepsilon > 0$ (assuming $O(|V|^{1/4 - \varepsilon})$-inapprox of Densest k-subgraph)	Solvable in factor-critical graphs, Exact algorithm in graphs G with GED (B, C, D) in time $2^{|C|}\text{poly}(|V(G)|)$, $\min\{OPT, \sqrt{|V(G)|}\}$-approx in graphs G whose GED (B, C, D) has no trivial components in $G[B]$

$$\nu(G, w) := \max \left\{ \sum_{e \in E} w_e x_e : \sum_{e \in \delta(v)} x_e \leq 1 \; \forall \, v \in V, \; x \in \mathbb{Z}_+^E \right\}. \qquad (1)$$

Definition 1 A *cooperative matching game instance* is an edge-weighted graph (G, w). The value of a coalition S is $\nu(G[S], w)$ and the value of the game is $\nu(G, w)$. The *core* of an instance (G, w) consists of allocations $y \in \mathbb{R}_+^V$ satisfying $\sum_{u \in V} y_u = \nu(G, w)$ and $\sum_{u \in S} y_u \geq \nu(G[S], w)$ for every $S \subseteq V$. A weighted graph (G, w) is defined to be *stable* if its core is non-empty.

Relaxing the integrality constraints in $\nu(G, w)$, we obtain the maximum weight fractional matching linear program:

$$\nu_f(G, w) := \max \left\{ \sum_{e \in E} w_e x_e : \sum_{e \in \delta(v)} x_e \leq 1 \; \forall \, v \in V, \; x_e \geq 0 \; \forall \, e \in E \right\}. \qquad (2)$$

The dual LP formulates the minimum fractional w-vertex cover:

$$\tau_f(G, w) := \min \left\{ \sum_{u \in V} y_u : y_u + y_v \geq w_{\{u,v\}} \ \forall \ \{u, v\} \in E, \ y_v \geq 0 \ \forall \ v \in V \right\}.$$
(3)

Imposing integrality requirements on the variables of the dual LP formulates the minimum w-vertex cover:

$$\tau(G, w) := \min \left\{ \sum_{u \in V} y_u : y_u + y_v \geq w_{\{u,v\}} \ \forall \ \{u, v\} \in E, \ y_v \in \mathbb{Z}_+ \ \forall \ v \in V \right\}.$$
(4)

The fractional problems are relaxations of the integral formulations. Further, by LP duality, the weight of a maximum fractional matching is equal to the weight of a minimum fractional w-vertex cover. Hence, we have the following relation:

$$\nu(G, w) \leq \nu_f(G, w) = \tau_f(G, w) \leq \tau(G, w).$$
(5)

We will use $\sigma(G, w) := \nu_f(G, w) - \nu(G, w)$ to denote the additive integrality gap of the fractional matching LP. It is well-known that every basic feasible solution to $\nu_f(G, w)$ is half-integral and moreover, the edges with half-integral components induce vertex disjoint odd cycles [5].

The rest of the treatise, except Sects. 5 and 6, will focus on uniform weight graphs, i.e., $w = \mathbb{1}$. When dealing with uniform weight graphs, we will omit the argument w for brevity (even from the ν, τ and σ notations).

2   Characterizing Stable Graphs

Before addressing the stabilization problems, we will discuss alternative and efficient characterizations of stable graphs and identify some well-known subfamilies of stable graphs. We will also illustrate the significance of stable graphs in the contexts of optimization and graph theory.

LP-based Characterization. The following alternative characterization of stable graphs is well-known.

Theorem 1 *[14, 31] A graph $G = (V, E)$ with edge weights $w : E \to \mathbb{R}_+$ is stable iff $\nu(G, w) = \nu_f(G, w)$.*

Theorem 1 implies that the algorithmic problem of verifying whether a given edge-weighted graph (G, w) is stable is solvable in polynomial time: we can solve the integer program $\nu(G, w)$ using Edmonds' max weight matching algorithm [16, 17], solve the linear program $\nu_f(G, w)$ using the Ellipsoid algorithm [25] and verify if the two values are equal. Further, *a witness of stability* for a stable graph is a matching M and a feasible fractional w-vertex cover y such that y satisfies complementary slackness conditions with the indicator vector of M. We will refer to such a witness of stability by the tuple (M, y).

In the terminology of discrete optimization, the *additive integrality gap of an instance for an LP* is the magnitude of the difference between the objective values of the optimum integral solution and the optimum LP solution. Using this terminology, an edge weighted graph (G, w) is stable iff the additive integrality gap of the instance (G, w) for the fractional matching LP is zero. Thus, the goal of stabilization is to modify the given instance (G, w) so that the additive integrality gap of the resulting instance (G', w') for the fractional matching LP is zero.

By Egerváry's theorem, if G is bipartite and $w \in \mathbb{Z}_+$, then $\nu(G, w) = \tau(G, w)$. Thus, bipartite graphs with integral edge weights are stable. In particular, unit-weight graphs with $\nu(G) = \tau(G)$ are known as *König-Egerváry graphs* [32, 33, 42] and by Theorem 1, König-Egerváry graphs are stable. Considering unit-weight graphs, we have the following strict containment relation: (see Figs. 1 and 2 illustrating strict containment):

$$\text{Bipartite} \subsetneq \text{König-Egerváry} \subsetneq \text{Stable}$$

Thus, stabilization closely resembles the goal of modifying a given graph to convert it into a König-Egerváry graph or a bipartite graph, both of which have been well-studied [1, 35]. While bipartite graphs are monotonic, i.e., closed under subgraphs, König-Egerváry graphs and stable graphs are not monotonic. For this reason, standard techniques to address graph modification problems to attain a monotonic property, e.g., [3], are not applicable for stabilization.

Graph-theoretic Characterization. We now mention equivalent graph-theoretic characterizations of unit-weight stable graphs. We will need some terminology

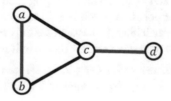

Fig. 1 A König-Egerváry graph that is not bipartite: $\nu(G) = 2$ and $S = \{a, c\}$ is a minimum vertex cover. However, the graph contains an odd cycle and hence is not bipartite

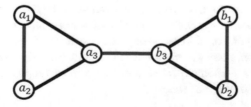

Fig. 2 A stable graph that is not König-Egerváry: $\nu(G) = 3$ and $\nu_f(G) = \tau_f(G) = 3$, since $y_{a_i} := 1/2$, $y_{b_i} := 1/2$ for all $i \in \{1, 2, 3\}$ is a feasible fractional vertex cover. However, $\tau(G) \geq 4$ since any feasible vertex cover S should have at least two vertices from $\{a_1, a_2, a_3\}$ as well as $\{b_1, b_2, b_3\}$

related to matchings. Let M be a matching in a graph $G = (V, E)$. A vertex $u \in V$ is said to be M-exposed if none of the edges in M contain u. A path is called M-*alternating* if it alternates edges from M and $E \setminus M$. An odd cycle of length $2k + 1$ in which exactly k edges are in M, for any positive integer k, is known as an M-*blossom*. The M-exposed vertex of an M-blossom is known as the *base vertex* of the M-blossom. An M-*flower* consists of an M-blossom with an even-length M-alternating path from the base vertex of the blossom to an M-exposed vertex. A matching in G is said to be a maximum matching if it has the maximum cardinality. A vertex $u \in V$ is said to be *inessential* if there exists a maximum matching in G that exposes u and is said to be essential otherwise. The *Gallai-Edmonds decomposition* (GED) [17, 21, 22] of a graph $G = (V, E)$ is a three partitioning (B, C, D) of the vertex set, where B is the set of inessential vertices, $C := N_G(B)$ and $D := V \setminus (B \cup C)$.

Theorem 2 *[4, 31, 43] Let $G = (V, E)$ be a graph with GED (B, C, D). The following are equivalent:*

1. *G is stable.*
2. *The induced subgraph $G[B]$ has no edges, i.e., the set of inessential vertices of G forms an independent set.*
3. *G contains no M-flower for every maximum matching M.*

Moreover, if G is unstable, then G contains an M-flower for every maximum matching M.

The Gallai-Edmonds decomposition (GED) and its properties will play an important role in stabilization. The GED of a graph is unique and can be found efficiently [17]. The GED of a graph contains valuable information that can be used to obtain optimal solutions to $\nu(G)$, $\nu_f(G)$ and $\tau_f(G)$. The following theorem summarizes some of these properties for our purposes. A graph is said to be *factor-critical* if for every vertex $u \in V$, there exists a matching M that exposes only u. A connected component is said to be *trivial* if it has only one vertex, otherwise it is said to be *non-trivial*.

Theorem 3 *(e.g. see Schrijver [39]) Let (B, C, D) denote the GED of a graph $G = (V, E)$ with B_1 being the subset of vertices of B which induce trivial components in $G[B]$ and $B_3 := B \setminus B_1$.*

1. *Each connected component in $G[B]$ is factor-critical.*
2. *Every maximum matching M in G contains a perfect matching in $G[D]$ and matches each vertex in C to distinct components in $G[B]$.*
3. *For every maximum matching M in G and every connected component K in $G[B]$, either (i) M exposes one vertex from K and has no edges leaving K or (ii) M does not expose any vertex from K and has exactly one edge leaving K.*
4. *The subset C is a choice of U that achieves the minimum in the Tutte-Berge formula for maximum matching,*

$$\nu(G) = \frac{1}{2} \min_{U \subseteq V} \{|V| + |U| - \text{number of odd-sized components in } G \setminus U\}$$

and hence C is known as the Tutte set *of G.*

5. Let M be a maximum matching that also matches the largest number of B_1 vertices. Then M exposes $2\sigma(G)$ vertices from B_3.
6. Let Y be a minimum vertex cover in the bipartite graph $H := (B_1 \cup C, \delta_G(B_1))$. Then, $y \in \mathbb{R}_+^V$ defined as follows is a minimum fractional vertex cover in G:

$$y(v) := \begin{cases} 1 & \text{if } v \in C \cap Y, \\ 0 & \text{if } v \in B_1 \setminus Y, \\ 1/2 & \text{otherwise.} \end{cases}$$

For notational convenience we will denote the GED of a graph by (B, C, D) as well as $(B = (B_1, B_3), C, D)$ where B_1 and B_3 are as defined in Theorem 3 above.

3 Edge Modifications

In this section, we will focus on stabilizing a given graph by edge-deletion and edge-addition.

3.1 Edge Deletion

In the stabilization problem by edge deletion (min-EDGE-DEL), the input is a graph $G = (V, E)$, and the goal is to find a minimum cardinality subset F of edges to delete to stabilize the graph. This is equivalent to enabling non-empty core for a given cooperative matching game instance by blocking the smallest number of relationships. There is always a feasible solution to min-EDGE-DEL since removing all edges of G will give a stable graph. Moreover, enabling non-empty core to a matching game instance by blocking the smallest number of relationships *does not* decrease the value of the game:

Theorem 4 *[10] For every optimum solution F^* to min-EDGE-DEL in G, we have $\nu(G \setminus F^*) = \nu(G)$.*

We sketch an algorithmic proof. Let M be a maximum matching in G with minimum $|M \cap F^*|$. Suppose $M \cap F^* \neq \emptyset$ for the sake of contradiction. Let $H := (G \setminus F^*) + M$. Since M is a maximum matching in H, there exists an M-flower R in H starting at an M-exposed vertex w. If the M-flower R contains an edge from F^*, then we can switch M along an even M-alternating path to obtain a maximum matching M' with fewer edges from F^*. Hence, R is an $(M \setminus F^*)$-flower in H. This $(M \setminus F^*)$-flower R is also present in $G \setminus F^*$. Since $G \setminus F^*$ is stable, by the characterization of stable graphs in Theorem 2, the matching $M \setminus F^*$ is not a maximum matching in $G \setminus F^*$.

Now, consider executing Edmonds' maximum matching algorithm on the graph $G \setminus F$ using $M \setminus F^*$ as the initial matching and construct an $(M \setminus F^*)$-*alternating* *tree* starting from the $(M \setminus F^*)$-exposed vertex w. Then, we will find either an $(M \setminus F^*)$-augmenting path P starting at w or a *frustrated tree* rooted at w. If we find an $(M \setminus F^*)$-augmenting path P starting at w, then the other end-vertex of P should be adjacent to an edge $f \in M \cap F^*$ since M is a maximum matching in G. Switching M along $P + f$ gives another maximum matching M' in G with $|M' \cap F^*| < |M \cap F^*|$, a contradiction. If we find a frustrated tree rooted at w, say with vertex set T, then the final maximum matching M' in $G \setminus F^*$ identified by Edmonds' algorithm has $M' \cap E[T] = M \cap E[T]$. Thus, the M-flower R is also an M'-flower, a contradiction to the stability of $G \setminus F^*$.

Hardness. min-EDGE-DEL is NP-hard by a reduction from the Vertex-Cover problem. In fact, there is an approximation preserving reduction (using a gadget) from the Vertex-Cover problem to min-EDGE-DEL in factor-critical graphs [10]. Consequently, min-EDGE-DEL has no efficient $(2 - \varepsilon)$-approximation algorithm for any $\varepsilon > 0$ even in factor-critical graphs assuming the Unique Games Conjecture [30].

 Lower bound. To design approximation algorithms, we first need a lower bound on the cardinality of the optimum solution $|F^*|$. Note that removing an edge can decrease the value of the minimum fractional vertex cover by atmost one. Consequently, removal of an edge can decrease the additive integrality gap for the fractional matching LP by at most one. Now, consider an arbitrary ordering of the edges in the optimum solution F^* and let F_i^* denote the first i edges according to this ordering and $F_0^* = \emptyset$. Then,

$$\sigma(G) - \sigma(G \setminus F^*) = \sum_{i=1}^{|F^*|} \left(\sigma(G - F_i^*) - \sigma(G - F_{i-1}^*) \right) \le |F^*|.$$

Since $\sigma(G \setminus F^*) = 0$, we have that $|F^*| \ge \sigma(G)$. Properties of GED can be used to tighten this lower bound to $|F^*| \ge 2\sigma(G)$ [10]. We will later see GED in action to obtain this tight lower bound in Sect. 4.1.

Approximations. On the approximation side, min-EDGE-DEL has a 4ω-approximation in ω-sparse graphs and a 2-approximation in regular graphs [10]. A graph is said to be ω-sparse if $|E[S]| \le \omega|S|$ for all $S \subseteq V$. We discuss the approximation in ω-sparse graphs. The core idea lies in the following lemma that follows from classic results on the structure of extreme point solutions to $\nu_f(G)$ and $\tau_f(G)$ [4, 43].

Lemma 1 *If G is an unstable ω-sparse graph, then there exists an efficient algorithm to find a subset L of edges such that (i) $|L| \le 4\omega$, (ii) $\nu(G \setminus L) = \nu(G)$, and (iii) $\nu_f(G \setminus L) \le \nu_f(G) - 1/2$.*

By the above lemma, we can find a small-sized subset L of edges such that the removal of L preserves the cardinality of the maximum matching while decreasing the additive integrality gap for the fractional matching LP by at least $1/2$. Thus, an algorithm to stabilize by edge deletion is to repeatedly apply the lemma until the graph becomes stable. In each application, we remove 4ω edges. The total number of iterations is not more than $2\sigma(G)$ since the additive integrality gap reduces by at least $1/2$ in each iteration. Hence the total number of edges removed is at most $8\omega(\nu_f(G) - \nu(G))$. Combining this with the lower bound mentioned above shows a 4ω-approximation.

An LP-relaxation. The improved 2-approximation for d-regular graphs is obtained by strengthening Lemma 1 to argue that $|L| \le d$ and tightening the lower bound using an LP-relaxation of min-EDGE-DEL. We now discuss this LP. Consider the following IP formulation of the min-EDGE-DEL problem, which we denote as $IP_{\text{edge-del}}$.

$$\min \left\{ \sum_{e \in E} z_e : y_u + y_v + z_{\{u,v\}} \ge 1 \; \forall \, \{u, v\} \in E, \sum_{u \in V} y_u = \nu(G), y \in \mathbb{R}_+^V, z \in \{0, 1\}^E \right\}.$$

The variable $z_{\{u,v\}}$ indicates if the edge $\{u, v\}$ is to be deleted while the y variables correspond to a fractional vertex cover that will be a witness of stability after deleting the support of z. The validity of the formulation follows from Theorem 4. Considering the LP-relaxation of $IP_{\text{edge-del}}$, we obtain the following dual:

$$\max \left\{ \sum_{e \in E} \alpha_e - \gamma\nu(G) : \sum_{e \in \delta(u)} \alpha_e \le \gamma \; \forall \, u \in V, 0 \le \alpha_e \le 1 \; \forall \, e \in E \right\}$$

In d-regular graphs, the solution $(\gamma = d, \alpha_e = 1 \; \forall \, e \in E)$ is a feasible dual solution and thus its objective value of $(d/2)\sigma(G)$ is a lower bound on the primal optimum value and in turn on $|F^*|$. Bock et al. [10] exhibit an example of a graph (non-regular) that demonstrates that the integrality gap of the LP-relaxation of $IP_{\text{edge-del}}$ (in fact, a stronger LP with more valid constraints) is $\Omega(|V|)$.

Open Problem 1. What is the approximability of min-EDGE-DEL in factor-critical graphs? It is impossible to obtain a $(2 - \varepsilon)$-approximation subject to the unique games conjecture as mentioned above. On the other hand, the main bottle-neck in designing approximation algorithms is the lack of a good lower bound. The current known lower bound arguments (discussed above) only show that at least one edge needs to be deleted from a factor-critical graphs to stabilize it. However, there exist factor-critical graphs from which $\Omega(|V|)$ edges need to be deleted for stabilizing [10].

Open Problem 2. In the max-EDGE-SUBGRAPH problem, the input is a graph $G = (V, E)$ and the goal is to find a subgraph $F \subseteq E$ with the largest number of edges such that (V, F) is stable. Even though max-EDGE-SUBGRAPH is equivalent to min-EDGE-DEL from the perspective of exact solvability, they differ in the approximability (similar to min-BIPARTIZATION and max-CUT). max-EDGE-SUBGRAPH is NP-hard and we can obtain a $(5/3)$-approximation: Mishra et al. [35] give an algorithm to find a subset of edges of cardinality at least $5|E|/3$ such that the induced subgraph is König-Egerváry. However, stable graphs are a super-family of König-Egerváry and it might be possible to improve on this approximation factor. What is the approximability of max-EDGE-SUBGRAPH?

3.1.1 Edge Deletion to Stabilize a Given Matching

In cooperative matching game instances, the central network authority might be interested in stabilizing so that a chosen maximum matching is preserved. Formally, in min-EDGE-DEL-M-STAB, the input is a graph $G = (V, E)$ and a maximum matching M in G, while the goal is to find a subset F of edges that is disjoint from M whose deletion stabilizes the graph. This is to be viewed as stabilizing a particular matching M in the graph. There is always a feasible solution to min-EDGE-DEL-M-STAB since we may delete all non-matching edges.

Connections to min-EDGE-DEL. Edge deletion to stabilize a chosen matching is also of interest as an intermediate problem towards the approximability of min-EDGE-DEL. By Theorem 4, we know that some maximum matching M survives the deletion of the optimal solution to min-EDGE-DEL. If we can find such a matching M (that survives the deletion of the optimal solution to min-EDGE-DEL), then we may simply solve (approximate) min-EDGE-DEL-M-STAB with respect to M to obtain an (approximately) optimal solution for min-EDGE-DEL.

We currently do not know how to identify a matching M that survives the deletion of the optimum solution to min-EDGE-DEL. It is known that not every maximum matching may survive the deletion of the optimum solution to min-EDGE-DEL. Bock et al. [10] show an example with two matchings M and M' where the optimum to min-EDGE-DEL-M-STAB for matchings M and M' differ by a factor of $\Omega(|V|)$, so arbitrary choices of maximum matching M to stabilize will not lead to good approximations for min-EDGE-DEL.

Approximability. min-EDGE-DEL-M-STAB is NP-hard in weighted graphs [8]. For unit-weight graphs, we know tight approximability results for min-EDGE-DEL-M-STAB. By an approximation preserving reduction from the Vertex Cover problem [10], min-EDGE-DEL-M-STAB is NP-hard and has no efficient $(2 - \epsilon)$-approximation algorithm for any $\varepsilon > 0$ subject to Unique Games Conjecture even in factor-critical graphs. On the other hand, we can obtain a 2-approximation by an LP-based algorithm [10].

We briefly describe this LP-based algorithm. The LP is a modification of the LP-relaxation of $IP_{\text{edge-del}}$ where we impose the complementary slackness conditions with M explicitly:

$$\min \sum_{e \in E} z_e$$
$$y_u + y_v = 1 \ \forall \ \{u, v\} \in M$$
$$y_u + y_v + z_{\{u,v\}} \geq 1 \ \forall \ \{u, v\} \in E \setminus M, \ u, v \in V(M)$$
$$y_v + z_{\{u,v\}} \geq 1 \ \forall \ \{u, v\} \in E \setminus M, \ v \in V(M), \ u \notin V(M)$$
$$y, z \geq 0.$$

If G is bipartite, then the constraint matrix of the above LP is *totally unimodular* and hence all extreme point optimum solutions to the LP are integral. If G is non-bipartite, then a standard construction transforms the graph to a bipartite graph whose integral optimum solution can be used to obtain a 2-approximate solution for G [10].

3.2 Edge Addition

In the stabilization problem by edge addition (min-EDGE-ADD), the input is a graph $G = (V, E)$, and the goal is to find a minimum cardinality subset $F \subseteq \binom{|V|}{2} \setminus E$ of non-edges of G to add to stabilize the graph. This is equivalent to enabling non-empty core for a given cooperative matching game instance by introducing the smallest number of new relationships.

Feasibility. Note that min-EDGE-ADD may not have a feasible solution: For instance, consider a triangle which has $\nu_f(G) = 3/2$ while $\nu(G) = 1$ and there are no more non-edges to add. More generally, we have the following family of infeasible instances.

Lemma 2 *[27] If $|V|$ is odd and $\nu_f(G) = |V|/2$, then min-EDGE-ADD has no feasible solution.*

The lemma follows from the following observation: Since $\nu_f(G) = |V|/2 = \tau_f(G)$, the addition of any new non-edges to G will not decrease the optimum value of the fractional vertex cover and hence the fractional matching value will be at least $|V|/2$. On the other hand, since $|V|$ is odd, the size of the maximum matching after the addition of any new edges to G can be at most $(|V| - 1)/2$. Thus, adding new edges can at best reduce the additive integrality gap for the fractional matching LP to $1/2$ but not to zero.

Efficient Solvability. In fact, the family of graphs satisfying the hypothesis of Lemma 2 are the only unstable graphs that cannot be stabilized by adding edges. The remaining graphs can be stabilized by adding edges.

Theorem 5 *[27] If either $|V|$ is even or $\nu_f(G) < |V|/2$, then an optimum solution to min-EDGE-ADD can be found efficiently and its cardinality is equal to $\lceil \sigma(G) \rceil$.*

We sketch a proof assuming $2\sigma(G)$ is odd (the other case is similar). Since $2\sigma(G)$ is odd, we have $\nu_f(G) < |V|/2$ (otherwise, $|V| = 2\nu_f(G) = 2|M| + 2\sigma(G)$ by Theorem 3 and hence, $|V|$ is odd, a contradiction). A lower bound of $\lceil \sigma(G) \rceil$ on the size of the optimal solution follows from the observation that the addition of a single edge can decrease the additive integrality gap for the fractional matching LP by at most one (similar to the lower bound in Sect. 3.1).

Next we find a collection of $\lceil \sigma(G) \rceil$ edges whose addition stabilizes the graph. For this, consider the GED $(B = (B_1, B_3), C, D)$ of G and a maximum matching M that matches the largest number of B_1 vertices. By Theorem 3, the matching M exposes $2\sigma(G)$ vertices from B_3 with at most one from each component of $G[B_3]$. If there is no M-exposed vertex in B_1, then $|V| = 2|M| + 2\sigma(G) = 2\nu_f(G)$ contradicting $\nu_f(G) < |V|/2$. Let s be an M-exposed vertex from B_1. Let F^* denote an arbitrary pairing of the exposed nodes from B_3 and s. Then $|F^*| = \lceil \sigma(G) \rceil$. We will prove that $G + F^*$ is stable by showing that $\sigma(G + F^*) = 0$. For the integral optimum, we have $\nu(G + F^*) \geq \nu(G) + |F^*|$ since $M \cup F^*$ is a matching in $G + F^*$. It remains to bound the fractional optimum $\nu_f(G + F^*)$. Consider the fractional vertex cover y of G obtained using GED (Theorem 3). Increase $y(s)$ by $1/2$. Now, all end vertices u of F^* have $y_u \geq 1/2$. So y is also a feasible fractional vertex cover in $G + F^*$ and $\tau_f(G + F^*) \leq \tau_f(G) + 1/2$. Hence $\sigma(G + F^*) = \nu_f(G + F^*) - \nu(G + F^*) \leq \tau_f(G + F^*) - \nu(G) - |F^*| \leq \tau_f(G) + 1/2 - \nu(G) - |F^*| = 0$.

As a consequence of Theorem 5, enabling non-empty core to a matching game instance G by introducing the smallest number of new relationships will increase the value of the game by exactly $\lceil \sigma(G) \rceil$.

Polyhedral Description. The efficient solvability and the tight characterization of the optimal solution to min-EDGE-ADD raises the question of whether there exists a polyhedral description of the characteristic vectors of non-edges of a graph whose addition stabilizes the graph. This is unlikely to exist since the following min-Cost-EDGE-ADD problem is NP-hard [27]: In the min-Cost-EDGE-ADD problem, we are given a graph $G = (V, E)$ with costs $c : \binom{|V|}{2} \setminus E \to \mathbb{R}_+$ on the non-edges and the goal is to find a subset $F \subseteq \binom{|V|}{2} \setminus E$ of minimum cost whose addition stabilizes the graph.

Open Problem. What is the approximability of min-Cost-EDGE-ADD problem? We only know that the problem is NP-hard [27].

4 Vertex Modifications

In this section, we will focus on stabilizing a given graph by vertex-deletion and vertex-addition.

4.1 Vertex Deletion

In the stabilization problem by vertex deletion (min-VERTEX-DEL), the input is a graph $G = (V, E)$, and the goal is to find a minimum cardinality subset S of vertices whose removal stabilizes the graph. This is equivalent to enabling non-empty core for a given cooperative matching game instance by blocking the smallest number of players. There is always a feasible solution to min-VERTEX-DEL since removing all but one vertex will give a stable graph. Enabling non-empty core for a matching game instance by blocking the smallest number of players does not decrease the value of the game (similar to min-EDGE-DEL):

Theorem 6 *[2] For every optimum solution S^* to min-VERTEX-DEL in G, we have*
$\nu(G \setminus S^*) = \nu(G)$.

Efficient Solvability. In contrast to min-EDGE-DEL, stabilization by minimum
vertex deletion is solvable efficiently.

Theorem 7 *[2, 27] An optimum solution to min-VERTEX-DEL can be found effi-*
ciently and its cardinality is equal to $2\sigma(G)$.

We first outline the lower bound using the properties of GED mentioned in Theorem
3. Let S^* be an optimum, $H := G \setminus S^*$, $(B = (B_1, B_3), C, D)$ be the GED of G and
M be a maximum matching that matches the largest number of B_1 vertices. Then
M exposes one vertex from $2\sigma(G)$ components in $G[B_3]$. Let $K_1, \ldots, K_r$ be these
components.

For each such component K_i, we claim that at least one vertex $u_i \in V(K_i)$
becomes essential in H: otherwise, every vertex $u \in V(K_i) \setminus S^*$ is inessential in
H. If $|S^* \cap V(K_i)| \geq 1$, then every maximum matching N in H necessarily exposes
two vertices in $V(K_i)$ and hence by Property 3 of GED (Theorem 3) applied to G,
we have that N is not a maximum matching in G. Hence, $\nu(H) = |N| < \nu(G)$, a
contradiction to Theorem 6. If $S^* \cap V(K_i) = \emptyset$, then H is unstable, a contradiction.

Since u_i is essential in H, a maximum matching N in H will match u_i. We may
assume without loss of generality that u_i is M-exposed. By Theorem 6, the graph
$M \triangle N$ is a disjoint union of even paths and even cycles. The end-vertices of the paths
that start at u_i must necessarily be in S^*, otherwise we may switch the matching N
along this path to obtain a maximum matching N' in H, where u_i is N'-exposed,
contradicting that u_i is essential in H.

The above proof technique is essentially a charging argument: for each M-exposed
component in $G[B_3]$, there is a unique vertex in S^*. Ito et al. [27] present an alternative
proof of the lower bound based on GED, but without using Theorem 6.

For the upper bound, consider a maximum matching M that matches the largest
number of B_1 vertices. Then M exposes one vertex from $2\sigma(G)$ components in
$G[B_3]$. For each such component, pick an arbitrary vertex into S^*. Then $|S^*| =
2\sigma(G)$ and $\nu(G - S^*) = \nu(G)$. In order to show that $G - S^*$ is stable, we will bound
the fractional optimum value $\nu_f(G - S^*)$. Consider the fractional vertex cover y
of G obtained using GED (Theorem 3). For each vertex $u \in S^*$, we have $y_u =
1/2$. Now projecting y to the remaining graph $G - S^*$ gives a feasible fractional
vertex cover in $G - S^*$ with value $\tau_f(G) - |S^*|/2$. Hence, $\sigma(G - S^*) \leq \tau_f(G) -
|S^*|/2 - \nu(G) = 0$.

4.2 Min Cost Vertex Deletion

The efficient solvability and the tight characterization of the optimal solution to
min-VERTEX-DEL raises the next natural question of whether there exists a poly-
hedral description of the characteristic vectors of vertices whose deletion stabilizes

the graph. However this is unlikely to exist since the minimum cost version of the stabilization by vertex deletion is NP-hard. We elaborate on the approximability of this problem in this section.

In the stabilization problem by min cost vertex deletion (min-Cost-VERTEX-DEL), the input is a graph $G = (V, E)$ with vertex-costs $c : V \to \mathbb{R}_+$, and the goal is to find a subset S of vertices of minimum cost $\sum_{u \in S} c_u$ whose removal stabilizes the graph. This is equivalent to enabling non-empty core for a given cooperative matching game instance by blocking the least cost set of players. The motivation is that not all players are equally important and hence the cost of blocking a player needs to be taken into account.

Approximation. min-Cost-VERTEX-DEL is NP-hard [2, 27] and has a $(|C| + 1)$-approximation [2], where (B, C, D) is the GED of the given graph G. We now discuss the approximation. Let $(B = (B_1, B_3), C, D)$ be the GED of the given graph G. The following structural properties of the optimal solution follow from Theorem 6.

Lemma 3 *Let S^* be an optimum to min-Cost-VERTEX-DEL. Then (i) S^* consists of only vertices in B, (ii) S^* contains at most one vertex from each component in $G[B]$ and (iii) if S^* contains a vertex from a component in $G[B]$, then that vertex is the least cost vertex in that component.*

The structural observations in Lemma 3 simplify the problem: replace the given instance G by contracting the non-trivial components in $G[B]$ and giving them a cost equal to the least cost vertex in the component. Let B_3 denote the vertex set of the contracted components. Delete D and the edges in $E[C]$. Denote the resulting bipartite graph as $G_b = (B \cup C, F)$. Now consider the following B_3-essentializer problem.

In the min-Cost-ESSENTIALIZER problem, we are given a bipartite graph $G_b = (B \cup C, F)$, with costs on the B-vertices, $c : B \to \mathbb{R}_+$ and a subset $B_3 \subseteq B$ of vertices. The goal is to find a minimum cost subset $S \subseteq B$ of vertices to delete such that every vertex in $B_3 \setminus S$ becomes essential in $G_b \setminus S$.

There is an approximation preserving reduction from min-Cost-VERTEX-DEL to min-Cost-ESSENTIALIZER using the above construction: if S is a feasible solution to min-Cost-ESSENTIALIZER such that $\nu(G_b \setminus S) = \nu(G)$, then it gives a feasible solution to min-Cost-VERTEX-DEL of the same cost (by mapping $u \in S$ to the least cost vertex in the component that was contracted to u to obtain G_b). So the goal boils down to finding an approximately optimum solution S to min-Cost-ESSENTIALIZER such that $\nu(G_b \setminus S) = \nu(G_b)$. Ahmadian et al. [2] give a $(|C| + 1)$-approximation for min-Cost-ESSENTIALIZER by an LP-based algorithm.

An LP-relaxation. The IP formulation for min-Cost-ESSENTIALIZER by Ahmadian et al. [2] differs notably from the IP formulations that have appeared in the stabilization literature. We discuss this formulation and the LP-based $(|C| + 1)$-approximation. For each $v \in B$, introduce two indicator variables: z_v indicates if

$v \in S$ (i.e., to be deleted) and y_v indicates if v will be essential in $G_b \setminus S$. For each $u \in C$, introduce x_u to indicate if u will be matched to inessential nodes in every maximum matching of $G_b \setminus S$.

$$\min \sum_{u \in B} c_u z_u$$

$$y_v + z_v \geq 1 \; \forall \, v \in B_3 \tag{6}$$

$$x_u + y_v + z_v \geq 1 \; \forall \, \{u, v\} \in F, \; u \in C, \; v \in B_1 \tag{7}$$

$$y(B) + x(C) = |C| \tag{8}$$

$$y(N_{G_b}(A)) \geq |A| - x(A) \; \forall \, A \subseteq C \tag{9}$$

$$y, z \in \{0, 1\}^B, \; x \in \{0, 1\}^C$$

Constraint (6) formulates that each $v \in B_3$ is either deleted or becomes essential. Constraint (7) formulates that if $v \in B_1$ is not deleted, then either v is essential or the neighbor u is a vertex that is matched to an inessential vertex in a maximum matching in $G_b \setminus S$. Constraint (8) formulates that for each vertex in C, either it is matched to an inessential vertex in a maximum matching in $G_b \setminus S$ or there exists a unique vertex in B that is essential. Constraint (9) formulates Hall's condition: there exists a matching between $C \setminus \text{Support}(x)$ and $\text{Support}(y)$.

The approximation algorithm is to find an extreme point optimum to the LP-relaxation of the above IP and perform threshold rounding: set $S := \{v : z_v \geq 1/(|C| + 1)\}$. The approximation factor follows immediately. Ahmadian et al. exploit the properties of the extreme point solution to argue that the solution S is indeed a B_3-essentializer. Recall that we obtain a solution for min-Cost-VERTEX-DEL only if $\nu(G_b \setminus S) = \nu(G_b)$. So, in order to ensure that this condition is satisfied, we repair the solution S without losing feasibility: repeatedly remove vertices from S and add back into the graph if this operation increases the cardinality of the maximum matching. Tighter valid inequalities or alternate algorithmic techniques are needed to improve on the approximation factor since the integrality gap of the LP is $\Omega(|C|)$ [2].

Open Problem 1. Can we design an approximation algorithm for min-Cost-VERTEX-DEL whose approximation factor is independent of the size of the Tutte set (e.g., a constant factor approximation)? A first step is to consider input instances where each vertex has only one of two possible costs. Even this case is NP-hard [2], but no constant-factor approximation is known.

Open Problem 2. In the max-Cost-VERTEX-SUBGRAPH, the input is a graph $G = (V, E)$ with vertex costs $c : V \to \mathbb{R}_+$ and the goal is to find a subset of vertices $U \subseteq V$ with maximum cost such that the induced subgraph $G[U]$ is stable. This is equivalent to min-Cost-VERTEX-DEL from the perspective of exact solvability but not approximability. Ahmadian et al. [2] give a 2-approximation following the above reduction to min-Cost-ESSENTIALIZER and based on the associated LP. What is the approximability of max-Cost-VERTEX-SUBGRAPH?

4.3   Vertex Addition

In the stabilization problem by vertex addition (min-VERTEX-ADD), the input is a
graph $G = (V, E)$, and the goal is to find a minimum cardinality subset S of vertices
along with some edges to V whose addition stabilizes the graph. This is equivalent to
enabling non-empty core for a given cooperative matching game instance by introducing the smallest number of new players with some relationships to the original
players. There is always a feasible solution to min-VERTEX-ADD since we may
pick an arbitrary maximum matching M and for each M-exposed vertex u, we can
add a new vertex v that is adjacent to u. In fact, min-VERTEX-ADD has a tight
characterization similar to min-VERTEX-DEL and min-EDGE-ADD.

Theorem 8 *[27] An optimum solution to min-VERTEX-ADD can be found efficiently
and its cardinality is equal to $2\sigma(G)$.*

The proof technique is similar to the one discussed for min-EDGE-ADD (Theorem
5). In particular, the algorithm adds exactly $2\sigma(G)$ vertices with exactly one edge
from each of these vertices, so the algorithm is also optimal if the goal is to minimize
the number of edges adjacent to the newly added vertices. As a consequence of
Theorem 8, optimal ways to enable non-empty core to a matching game instance G
by introducing new players will increase the profit of the game by exactly $2\sigma(G)$.

5   Edge Weight Addition

In the stabilization problem by edge weight addition (min-EDGE-WT-ADD), the
input is an edge-weighted graph $(G = (V, E), w')$. The goal is to find an increase in
the edge weights $w : E \to \mathbb{R}_+$ so that the resulting edge-weighted graph $(G, w' + w)$
is stable and moreover the total increase in edge-weights $\sum_{e \in E} w_e$ is minimized. This
is equivalent to enabling non-empty core for a given cooperative matching game
instance by minimally increasing the weights on the edges.

In understanding the complexity of min-EDGE-WT-ADD, it is imperative to first
address input graphs with uniform edge-weights, i.e., $w' = \mathbb{1}$. These are also the
input graphs discussed in the previous sections. In the rest of this section, we will
assume that the inputs are unit-weight graphs.

min-EDGE-WT-ADD is a continuous optimization problem as opposed to the
stabilization problems considered in Sects. 3 and 4 which are discrete optimization
problems. Note that min-EDGE-WT-ADD is not a continuous version of min-EDGE-
ADD since the edge weights are allowed to increase only on the edges of the given
graph.

There is always a feasible solution to min-EDGE-WT-ADD: consider a maximum
matching M in G and increase the weights on the matching edges by one unit, i.e.,
set $w(e)$ to be 1 for $e \in M$ and 0 for $e \in E \setminus M$. Then, the characteristic vector of
M is an optimum solution to $\nu(G, \mathbb{1} + w)$ while $y_v = 1$ for all vertices $v \in V(M)$ is

a feasible fractional $(\mathbb{1} + w)$-vertex cover. Since $\nu(G, \mathbb{1} + w) = \tau_f(G, \mathbb{1} + w)$, we have that (M, y) is a witness of stability for $(G, \mathbb{1} + w)$.

The following result shows that every optimal edge-weight addition w to stabilize a unit-weight graph preserves the number of matching edges in the maximum $(\mathbb{1} + w)$-weight matching. Thus, enabling non-empty core for a given uniform-weight cooperative matching game instance by minimally increasing the weights does not decrease the number of matching edges in the grand-coalition.

Theorem 9 *[12] For every optimal solution w^* to min-EDGE-WT-ADD, every maximum $(\mathbb{1} + w^*)$-weight matching M^* has the same cardinality as $\nu(G)$. Moreover, $w^*(e) = 0$ on all edges $e \in E \setminus M^*$ and $0 \le w^*(e) \le 1$ for all edges $e \in M^*$.*

Thus, a maximum cardinality matching becomes a maximum weight matching after stabilizing by minimum edge weight addition. The structural properties of the optimum given in Theorem 9 can be used to solve min-EDGE-WT-ADD in factor-critical graphs.

Efficient algorithm in factor-critical graphs. Let M^* denote the maximum weight matching after stabilizing a factor-critical graph G by minimum edge weight addition. By Theorem 9, M^* exposes exactly one vertex. By guessing this vertex $a \in V$, we can find an optimum solution to min-EDGE-WT-ADD as follows: Find a maximum cardinality matching M exposing a. Solve the minimum fractional vertex cover LP with the additional constraint that $y_a = 0$. Now, set $w_{\{u,v\}} = y_u + y_v - 1$ for each matching edge $\{u, v\} \in M$ and $w_e = 0$ for non-matching edges. It is a simple exercise to show that (M, y) is a witness of stability for $(G, \mathbb{1} + w)$.

We will show that w is an optimum. Consider an optimum solution w^* to min-EDGE-WT-ADD. Let (M^*, y^*) be a witness of stability for $(G, \mathbb{1} + w^*)$. By complementary slackness conditions, $y_a^* = 0$ and $w^*(\{u, v\}) = y_u^* + y_v^* - 1$ for matching edges $\{u, v\} \in M^*$. We have the lower bound by the following:

$$\sum_{e \in E} w^*(e) = \sum_{\{u,v\} \in M^*} y_u^* + y_v^* - 1 = \sum_{u \in V} y_u^* - |M^*| = \sum_{u \in V} y_u - |M| = \sum_{e \in E} w(e).$$

The above argument shows that the precise choice of the matching M^* does not influence the optimum edge weight addition in factor-critical graphs; instead, the vertex exposed by M^* completely determines the optimum edge weights.

Arbitrary graphs: reducing to discrete decision domain. The following stronger structural properties of the optimum solution are helpful in the investigation of the complexity of min-EDGE-WT-ADD for arbitrary graphs.

Theorem 10 *[12] There exists an optimal solution w^* to min-EDGE-WT-ADD that is half-integral with a witness (M^*, y^*) of stability of $(G, \mathbb{1} + w^*)$ such that $y^* \in \{0, 1/2, 1\}^V$ with Support(y^*) containing the Tutte set.*

Theorem 10 simplifies the continuous decision domain of min-EDGE-WT-ADD to the discrete decision domain: the goal is to determine the half-integral weight increase for each edge.

In [12], we take an alternative perspective of the decision domain: we show an efficient algorithm to find the optimum w^* if we know the values of y^* for the Tutte vertices (the algorithm crucially relies on the solvability of min-EDGE-WT-ADD in factor-critical graphs). On the one hand, this immediately implies an algorithm to solve min-EDGE-WT-ADD exactly in time that is exponential only in the size of the Tutte set. On the other hand, the algorithm suggests that the complexity of min-EDGE-WT-ADD essentially lies in deciding whether the vertex-cover value y_v^* on each Tutte vertex v is $1/2$ or 1. We use this alternative discrete perspective to reduce the SET-COVER problem to min-EDGE-WT-ADD in an approximation preserving fashion (up to constant-factors). Consequently, min-EDGE-WT-ADD is inapproximable to a factor better than $c \log |V|$ for some constant c [12]. This is the only known super-constant inapproximability result for any of the stabilization models.

Yet another discrete perspective shows that min-EDGE-WT-ADD is possibly inapproximable to a much larger factor than $c \log |V|$: Suppose the GED (B, C, D) has only non-trivial components in $G[B]$. Then, knowledge of the factor-critical components of $G[B]$ that are exposed by M^* is sufficient to solve the problem exactly. However, there is a reduction from DENSEST k-SUBGRAPH to the problem of determining the M^*-exposed factor-critical components of $G[B]$ and in turn to min-EDGE-WT-ADD [12]. In the DENSEST k-SUBGRAPH, we are given a graph $H = (U, F)$, a positive integer k and the goal is to find a subset S of k vertices such that the induced subgraph $H[S]$ has the largest number of edges. DENSEST k-SUBGRAPH is believed to be a difficult problem [29] and possibly inapproximable to a factor better than the current best known $O(|V|^{1/4})$ [7]. The reduction from DENSEST k-SUBGRAPH suggests that min-EDGE-WT-ADD is unlikely to be approximable to a factor better than $O(|V|^{1/16})$.

In the positive direction, we have an algorithm that attains a $\min\{OPT, \sqrt{|V|}\}$-approximation factor for min-EDGE-WT-ADD in graphs G whose GED (B, C, D) has only non-trivial components in $G[B]$ [12].

Open Problem. Can we obtain an $O(\sqrt{|V|})$-approximation for min-EDGE-WT-ADD? Currently, we have a $O(\sqrt{|V|})$-approximation for min-EDGE-WT-ADD only in graphs whose GED (B, C, D) has no trivial components in $G[B]$.

6   Further Open Problems

We conclude with further open problems related to alternative stabilization models and weighted graphs.

Unit-weight graphs. For unit-weight graphs, we have the following stabilization problem in addition to the open problems mentioned in Sects. 3, 4, and 5.

1. Similar to edge weight addition, we may also consider edge weight deletion (min-EDGE-WT-DEL). The input here is a graph $G = (V, E)$ with unit weights on the

edges and the goal is to reduce the weights on the edges $w : E \to \mathbb{R}_+$ so that the resulting weighted graph $(G, \mathbb{1} - w)$ is stable and moreover the total decrease in edge-weights $\sum_{e \in E} w_e$ is minimized. This is equivalent to enabling non-empty core for a given cooperative matching game instance by minimally penalising the profits on some of the relationships. The complexity of this problem is open and this is perhaps solvable efficiently by an appropriate LP-formulation. Note that min-EDGE-WT-DEL is the continuous version of min-EDGE-DEL.

Weighted graphs. Given the limited knowledge of stabilization in unit-weight graphs, the following problems are realistic goals towards stabilizing weighted graphs $(G = (V, E), w : E \to \mathbb{R}_+)$.

1. In the min-EDGE-ADD problem in weighted graphs, we are given weights w' : $\binom{V}{2} \setminus E \to \mathbb{R}_+$ on the non-edges and the goal is to find a minimum cardinality subset of edges to add so that the resulting graph becomes stable. If w and w' are unity, then we have a thorough understanding of min-EDGE-ADD while the complexity of min-EDGE-ADD for arbitrary weights w and w' is open.
2. In the min-VERTEX-DEL problem in weighted graphs, the goal is to delete the smallest subset of vertices from G so that the remaining graph, with the given weights w on the surviving edges, becomes stable. If w is unity, then min-VERTEX-DEL is solvable efficiently. What is the complexity of min-VERTEX-DEL in weighted graphs?

Acknowledgements I would like to thank my collaborators in [10, 12] and the authors of [2, 27] for sharing their insights and results at various conference venues. Thanks also to Chandra Chekuri, Britta Peis and Corinna Gottschalk for helpful suggestions on previous drafts of the article.

References

1. A. Agarwal, M. Charikar, K. Makarychev, Y. Makarychev, $O(\sqrt{\log n})$ approximation algorithms for min UnCut, min 2CNF deletion, and directed cut problems, in *Proceeding 37th Annual ACM Symposium Theory Computing* (2005), pp. 573–581,
2. S. Ahmadian, H. Hosseinzadeh, L. Sanitá, Stabilizing network bargaining games by blocking players, in *Proceedings Integer Programming and Combinatorial Optimization: 18th International Conference, IPCO 2016* (2016), pp. 164–177
3. N. Alon, A. Shapira, B. Sudakov, Additive approximation for edge-deletion problems, in *Proceedings 46th Annual IEEE Symposium Foundations Computer Science* (2005), pp. 419–428
4. E. Balas, Integer and fractional matchings. N.-Holl. Math. Stud. **59**, 1–13 (1981)
5. M.L. Balinski, On maximum matching, minimum covering and their connections, in *Proceedings of the Princeton Symposium on Mathematical Programming* (1970)
6. M.H. Bateni, M.T. Hajiaghayi, N. Immorlica, H. Mahini, The cooperative game theory foundations of network bargaining games, in *Proceeding 37th International Colloquium on Automata, Languages and Programming* (Springer, 2010), pp. 67–78
7. A. Bhaskara, M. Charikar, E. Chlamtac, U. Feige, A. Vijayaraghavan, Detecting high log-densities: an $O(n^{1/4})$ approximation for densest k-subgraph, in *Proceeding 42nd Annual ACM Symposium Theory Computing* (2010), pp. 201–210

8. P. Biró, M. Bomhoff, P. A. Golovach, W. Kern, D. Paulusma, Solutions for the stable roommates problem with payments, in *Graph-Theoretic Concepts in Computer Science*. Lecture Notes in Computer Science, vol. 7551 (2012), pp. 69–80

9. P. Biró, W. Kern, D. Paulusma, On solution concepts for matching games, in *Theory and Applications of Models of Computation* (Springer, 2010), pp. 117–127

10. A. Bock, K. Chandrasekaran, J. Könemann, B. Peis, L. Sanità, Finding small stabilizers for unstable graphs. Math. Program. **154**(1), 173–196 (2015)

11. G. Chalkiadakis, E. Elkind, M. Wooldridge, Computational aspects of cooperative game theory. Synth. Lect. Artif. Intell. Mach. Learn. **5**(6), 1–168 (2011)

12. K. Chandrasekaran, C. Gottschalk, J. Könemann, B. Peis, D. Schmand, A. Wierz, Additive stabilizers for unstable graphs (2016), arXiv: abs/1608.06797

13. W. Cook, W. Cunningham, W. Pulleyblank, A. Schrijver, *Combinatorial Optimization (Chapter 5)* (Wiley, 1998)

14. X. Deng, T. Ibaraki, H. Nagamochi, Algorithmic aspects of the core of combinatorial optimization games. Math. Oper. Res. **24**(3), 751–766 (1999)

15. D. Easley, J. Kleinberg, *Networks, Crowds, and Markets: Reasoning About a Highly Connected World* (Cambridge University Press, 2010)

16. J. Edmonds, Maximum matching and a polyhedron with 0, 1-vertices. J. Res. Natl. Bur. Stand. B **69**(1965), 125–130 (1965)

17. J. Edmonds, Paths, trees, and flowers. Can. J. Math. **17**(3), 449–467 (1965)

18. K. Eriksson, J. Karlander, Stable outcomes of the roommate game with transferable utility. Int. J. Game Theory **29**(4), 555–569 (2000)

19. L. Farczadi, K. Georgiou, J. Könemann, Network bargaining with general capacities, in *Proceeding 21st Annual European Symposium on Algorithms* (2013), pp. 433–444

20. D. Gale, L. Shapley, College admissions and the stability of marriage. Am. Math. Mon. **69**(1), 9–14 (1962)

21. T. Gallai, Kritische Graphen II. Magy. Tud. Akad. Mat. Kutató Int. Közl **8**, 373–395 (1963)

22. T. Gallai, Maximale Systeme unabhängiger Kanten [German]. Magy. Tud. Akad. Mat. Kutató Int. Közl **9**, 401–413 (1964)

23. D. Gillies, Solutions to general nonzero sum games. Ann. Math. Stud. **IV**(40), 47–85 (1959)

24. D. Granot, A note on the room-mates problem and a related revenue allocation problem. Manag. Sci. **30**, 633–643 (1984)

25. M. Grötschel, L. Lovász, A. Schrijver, Geometric algorithms and combinatorial optimization, in *Algorithms and Combinatorics*, vol. 2, 2nd edn. (Springer, 1993)

26. J.W. Hatfield, P.R. Milgrom, Matching with contracts. Am. Econ. Rev. **95**(4), 913–935 (2005)

27. T. Ito, N. Kakimura, N. Kamiyama, Y. Kobayashi, Y. Okamoto, Efficient stabilization of cooperative matching games, in *Proceedings of the 2016 International Conference on Autonomous Agents and Multiagent Systems, AAMAS 2016* (2016), pp. 41–49

28. M.O. Jackson, *Social and Economic Networks* (Princeton University Press, Princeton, 2008)

29. S. Khot, Ruling out PTAS for graph min-bisection, dense k-subgraph, and bipartite clique. SIAM J. Comput. **36**(4), 1025–1071 (2005)

30. S. Khot, O. Regev, Vertex cover might be hard to approximate to within 2- ε. J. Comput. Syst. Sci. **74**(3), 335–349 (2008)

31. J. Kleinberg, É. Tardos, Balanced outcomes in social exchange networks, in *Proceedings 40th Annual ACM Symposium Theory Computing* (2008), pp. 295–304

32. E. Korach, Flowers and trees in a ballet of K_4, or a finite basis characterization of non-König-Egerváry graphs. Technical Report No. 115 (IBM Israel Scientific Center, 1982)

33. E. Korach, T. Nguyen, B. Peis, Subgraph characterization of red/blue-split graph and König Egerváry graphs, in *Proceedings, ACM-SIAM Symposium on Discrete Algorithms* (2006), pp. 842–850

34. L. Lovász, M. Plummer, *Matching Theory* (North Holland, 1986)

35. S. Mishra, V. Raman, S. Saurabh, S. Sikdar, C.R. Subramanian, The complexity of König subgraph problems and above-guarantee vertex cover. Algorithmica **61**(4), 857–881 (2011)

36. J. Nash Jr, The bargaining problem. Econom. J. Econom. Soc. 155–162 (1950)

37. N. Nisan, T. Roughgarden, É. Tardos, V. Vazirani, *Algorithmic Game Theory* (Cambridge University Press, Cambridge, UK, 2007)
38. A. Schrijver, *Theory of Linear and Integer Programming* (Wiley, 1998)
39. A. Schrijver, *Combinatorial Optimization: Polyhedra and Efficiency*, vol. 24 (Springer, Berlin, Germany, 2003)
40. L.S. Shapley, M. Shubik, The assignment game I: The core. Int. J. Game Theory **1**(1), 111–130 (1971)
41. Y. Shoham, K. Leyton-Brown, *Multiagent Systems: Algorithmic, Game-Theoretic, and Logical Foundations* (Cambridge University Press, 2008)
42. F. Sterboul, A characterization of the graphs in which the transversal number equals the matching number. J. Comb. Theory Ser. B (1979), pp. 228–229
43. J.P. Uhry, Sur le problème du couplage maximal. Revue française d'automatique, d'informatique et de recherche opérationnelle. Recherche opérationnelle **9**(3), 13–20 (1975)
44. V. Vazirani, *Approximation Algorithms* (Springer Inc., New York, 2001)
45. D. West, *Introduction to Graph Theory* (Prentice Hall, 2001)
46. D. Williamson, D. Shmoys, *The Design of Approximation Algorithms* (Cambridge University Press, 2011)

Spider Covering Algorithms for Network Design Problems

Takuro Fukunaga

Abstract The spider covering framework was originally proposed by Klein and Ravi (1993) in their attempt to design a tight approximation algorithm for the node-weighted network design problem. In this framework, an algorithm constructs a solution by repeatedly choosing a low-density graph. The analysis based on this framework basically follows from the idea used in the analysis of the well-known greedy algorithm for the set cover problem. After Klein and Ravi, the framework has been sophisticated in a series of studies, and efficient approximation algorithms for numerous network design problems have been proposed. In this article, we survey these studies on the spider covering framework.

1 Introduction

Network design problems are optimization problems of constructing a small-weight network that connects given terminal nodes under some constraints. They include many fundamental combinatorial optimization problems such as the spanning tree, Steiner tree, and Steiner forest problems, as well as survivable network design problems (SNDPs). All of these problems have been studied actively in the area of combinatorial optimization, and those studies have provided numerous useful techniques for designing efficient approximation algorithms. Eventually, each of these techniques was applied to the other combinatorial optimization problems, indicating that network design problems are sources of useful ideas for combinatorial optimization.

In this article, we focus on the spider covering framework, which provides efficient algorithms for many network design problems. In this framework, an algorithm constructs a solution by repeatedly choosing a low-density graph. Here, the density of a graph is usually defined as its weight divided by the number of spanned terminals, and the solution is defined as the union of the chosen low-density graphs. Analysis of the algorithm basically follows from the idea used in the analysis of the well-known greedy algorithm for the set cover problem. In the set cover problem, the

T. Fukunaga (✉)
National Institute of Informatics, Hitotsubashi, Chiyoda-ku, Japan
e-mail: takuro@nii.ac.jp

© Springer Nature Singapore Pte Ltd. 2017
T. Fukunaga and K. Kawarabayashi (eds.), *Combinatorial Optimization and Graph Algorithms*, DOI 10.1007/978-981-10-6147-9_3

43

greedy algorithm repeats choosing a set of smallest density to cover all elements in an underlying set, and the key in its analysis is to show that there exists a set whose density is at most that of an optimal solution. To extend this analysis to network design problems, a type of graph called a spider is used in what is called the spider covering framework. Spiders are utilized because they have the following two useful properties: first, the minimum density of spiders can be bounded in terms of the density of an optimal solution; second, a graph whose density is at most the minimum density of spiders can be computed (exactly or approximately) in polynomial time. These two properties make it possible to extend the greedy algorithm for the set cover problem to many network design problems.

The spider covering framework was introduced by Klein and Ravi [1] for solving the node-weighted Steiner tree problem (its journal article version was published as [2]). While the Steiner tree problem admits a constant-factor approximation algorithm for edge-weighted graphs [3, 4], it is NP-hard to achieve an $o(\log |T|)$-approximation for the problem with node-weighted graphs because the problem includes the set cover problem, where T is the set of terminals in the given instance. Klein and Ravi presented an $O(\log |T|)$-approximation algorithm for the node-weighted Steiner tree problem by using the idea of spider covering. In the work of Klein and Ravi, a spider is defined as a tree that includes at most one node of degree larger than two. The second property of spiders given above easily follows from this definition. To prove the first property, Klein and Ravi proved the spider decomposition theorem, which claims that any Steiner tree can be decomposed into disjoint spiders.

Later, Guha et al. [5] provided an alternative proof for the first property. They bound the minimum density of spiders with respect to the optimal value of a linear programming (LP) relaxation of the problem. This alternative proof has several advantages over the original proof of Klein and Ravi. For example, it can be easily extended to other related problems such as the prize-collecting node-weighted Steiner tree problem.

After [1], their algorithm was extended to several network design problems, which increased the sophistication of the overall framework. One of the most important contributions in those follow-ups is an extension to the node-weighted SNDP, given by Nutov [6, 7]. He applied the idea to the problem of constructing a minimum node-weight network under constraints on connectivity. For this, the notion of spiders was extended in a nontrivial way.

In addition, Nutov [8] observed that the technique is useful for the network activation problem. This problem is an extension of the network design problems on node-weighted graphs. In the node-weighted graph, when we choose an edge as a part of a solution, the weights of both its end nodes are charged as costs. In the network activation problem, the task is to assign non-negative numbers to all nodes in the given graph. Each edge in the graph is associated with an activation function, and this function decides whether the edge is activated or not from the numbers assigned to its two end nodes. The objective is to minimize the sum of the numbers assigned to nodes under the constraint that the network formed by the activated edges satisfies some given requirements. The problem includes the node-weighted problem by setting the activation function so that it activates an edge if each of the end nodes is assigned a number at least its weight. Nutov [8] proved that the spider covering technique can be applied to the network activation problem. Fukunaga [9] pointed out a flaw in Nutov's claim for the case that the constraints are defined with regard to the node-connectivity, and presented a corrected version. The main idea in this correction is a new potential function that measures the progress of the greedy algorithm.

The purpose of the present article is to survey such progress made on the spider covering framework. Since it is impossible to cover all results given by the framework, we focus on Klein and Ravi's algorithm for the node-weighted Steiner tree, its LP-based analysis due to Guha et al., Nutov's algorithm for the node-weighted SNDP, and Fukunaga's algorithm for the node-connectivity network activation problem. We believe that these illustrate the essential ideas in the framework.

1.1 Organization

The rest of this article is organized as follows. Section 2 introduces the notation used throughout the article. Section 3 presents the Klein and Ravi's algorithm for the node-weighted Steiner tree problem and its analysis based on the spider decomposition theorem. Section 4 gives an alternative analysis on Klein and Ravi's algorithm given by Guha et al. Section 5 presents Nutov's algorithm for the node-weighted edge-connectivity SNDP, and Sect. 6 illustrates Fukunaga's algorithm for the node-connectivity network activation problem. Section 7 briefly describes other related work, and Sect. 8 concludes the article.

2 Preliminaries

We introduce the notation used throughout this article. We let $\mathbb{Z}_+$ and $\mathbb{R}_+$ denote the sets of non-negative integers and real numbers, respectively. For a positive integer i, we let $[i]$ denote the set of integers $\{1, \ldots, i\}$. For finite sets U and V with $U \subseteq V$ and a vector $f \in \mathbb{R}_+^U$, we represent $\sum_{i \in U} f(i)$ as $f(U)$.

Let $G = (V, E)$ be an undirected graph with a node set V and an edge set E. In this article, we focus on undirected graphs. Hence the graphs are undirected throughout the article unless otherwise stated. An unordered pair of elements u and v is denoted by $\{u, v\}$, and an ordered pair of them is denoted by (u, v) if u precedes v. An edge joining two nodes u and v is denoted by uv if it is clear from the context whether the edge is undirected or directed. For an edge set $F \subseteq E$, $V(F)$ denotes the set of end nodes of edges in F. In other words, $V(F) = \bigcup_{uv \in F} \{u, v\}$.

For $U \subseteq V$, $\delta_E(U)$ denotes the set of edges in E joining nodes in U with those in $V \setminus U$ (i.e., $\delta_E(U) = \{uv \in E : u \in U, v \notin U\}$), and $\Gamma_E(U)$ denotes the set of end nodes of edges in $\delta_E(U)$ that are not included in U (i.e., $\delta_E(U) = \{v \in V \setminus U : \exists uv \in E, u \in U\}$).

3 Spider Covering Algorithm for the Node-Weighted Steiner Tree

In this section, we introduce the standard definition of a spider, and demonstrate its usefulness by presenting Klein and Ravi's algorithm [2] for the node-weighted Steiner tree problem.

In the node-weighted Steiner tree problem, we are given an undirected graph $G = (V, E)$, node weights $w : V \to \mathbb{R}_+$, and a subset T of the node set V. Nodes in T are called *terminals* and those in $V \setminus T$ are called *Steiner nodes*. A tree is called a Steiner tree if it spans all terminals, as well as possibly spanning some of the Steiner nodes. We regard a tree F as a set of edges here. The weight of a tree F, denoted by $w(F)$, is defined as $\sum_{v \in V(F)} w(v)$. The objective of the problem is to find a minimum weight Steiner tree. In the following discussion, we assume without loss of generality that $w(v) = 0$ for each terminal $v \in T$.

The Steiner tree problem can also be defined with edge weights; the weight of a tree is defined as the sum of the weights of its edges. To distinguish this problem from the problem with node weights, let us call the problem with edge weights the *edge-weighted Steiner tree problem*. The edge-weighted Steiner tree problem can be reduced to the node-weighted Steiner tree problem by subdividing each edge by adding a new node associated with the weight of the original edge.

The edge-weighted Steiner tree problem has constant-factor approximation algorithms (the currently known best approximation factor is $\ln(4) + \epsilon < 1.39$ [3, 4]). In contrast, it is known that the node-weighted Steiner tree problem includes the set cover problem, which means that it is NP-hard to compute an $o(\log |T|)$-approximate solution for the node-weighted Steiner tree problem [2]. Klein and Ravi introduced the spider covering framework to present an $O(\log |T|)$-approximation algorithm for the node-weighted Steiner tree problem. Because of the above hardness, this approximation factor is tight up to a constant factor unless $P = NP$.

Now let us introduce Klein and Ravi's algorithm. The algorithm comprises iterations. In each iteration, it computes a connected subgraph spanning at least two terminals, and shrinks it into a single terminal. These iterations are repeated until the number of terminals reaches one, and then the algorithm outputs the set of original nodes corresponding to those included in the subgraph at each iteration. Since the number of terminals decreases by at least one in each iteration, the algorithm terminates after $O(|T|)$ iterations. Moreover, the node set output by the algorithm induces a connected subgraph spanning all terminals; namely, it induces a Steiner tree.

The crux of the algorithm is to bound the *density* of the subgraphs chosen in the iterations. For a subgraph S of G, let $T(S)$ denote the set of terminals included in $V(S)$. Then, the density of S is defined as the weight of S divided by the number of included terminals, i.e., $w(S)/|T(S)|$. Klein and Ravi proved that we can compute a subgraph of density $O(1) \cdot \text{OPT}/|T|$ in polynomial time, where OPT is the minimum weight of Steiner trees. By an analysis similar to that of the $O(\log n)$-approximation algorithm for the set cover, it can be seen that this indicates that the algorithm achieves $O(\log |T|)$-approximation.

A candidate for the low-density subgraph is a spider. Let us now define a spider.

Definition 1 A tree in the graph G is a spider if at most one node has degree larger than two, and all leaves are terminals. Such a unique node with degree larger than two is called the *head* of the spider. If the degrees of the nodes in the spider are at most two, then an arbitrary node in the spider is chosen to be the head.

We note that a spider can be decomposed into paths connecting the head and the leaves of the spider, and these paths do not share any node but the head. We do not regard a tree that consists of a single node as a spider. Hence a spider includes at least two terminals.

Ravi and Klein proved that there exists a low-density spider, as follows.

Theorem 1 *There exists a spider of density at most* $\text{OPT}/|T|$.

Indeed, Theorem 1 relies on the existence of a decomposition of a Steiner tree into node-disjoint spiders, which is described as follows.

Lemma 1 *Let F be a Steiner tree that spans at least two terminals. Then, there is a set of node-disjoint spiders included in F such that each terminal spanned by F is included in some spider.*

Proof We prove the theorem by induction on the number of nodes of degree greater than two in F. We can assume that all leaves of F are terminals by removing leaves that are not terminals. If F includes at most one node of degree greater than two, then F is a spider and the lemma holds. Suppose that there are at least two nodes of degree greater than two in F. Regard F as a rooted tree by choosing an arbitrary node as the root. Let v be a node of degree greater than two such that all of its descendants are of degree at most two. Then the subtree S_v rooted at v is a spider. Let F' denote the tree obtained from F by removing the subtree S_v. Since all leaves of F are terminals and there are at least two nodes of degree greater than two in F. F' includes at least two terminals. Hence, by induction, there are node-disjoint spiders in F' such that each terminal in F' is spanned by one of the spiders. The set of these spiders and S_v constitute the required decomposition of F. □

Lemma 1 implies Theorem 1. To see this, let F be a Steiner tree of weight OPT, and $S_1, \ldots, S_l$ be the spiders obtained from F by Lemma 1. Then $\sum_{i=1}^{l} w(S_i) \leq \text{OPT}$ because the spiders are node-disjoint, and $\sum_{i=1}^{l} |T(S_i)| = |T|$ because each terminal is included in one of the spiders. Hence $\min_{i=1}^{l} w(S_i)/|T(S_i)| \leq (\sum_{i=1}^{l} w(S_i))/ (\sum_{i=1}^{l} |T(S_i)|) \leq \text{OPT}/|T|$.

Despite this useful property of spiders, there are no polynomial-time algorithms for computing a low-density spider. Instead of spiders, we use a relaxation of spiders.

Lemma 2 *Let D be the minimum density of spiders in a graph. Then, there exists a polynomial-time algorithm to compute a subgraph of density at most D spanning at least two terminals.*

Proof Let S denote a spider of density D. Let $h \in V$ be the head of S, $t_1, \ldots, t_l$ be the leaves of S, and P_i denote the path connecting h and t_i in S for each $i \in [l]$. The definition of spiders indicates that all leaves are terminals and $l \geq 2$. If an inner node v of P_i is a terminal, then the subpath of P_i between t_i and v or the tree obtained from S by replacing P_i with its subpath between h and v gives another spider of density at most D. Thus, we assume without loss of generality that no inner nodes of the paths $P_1, \ldots, P_l$ are terminals. Then by this assumption, S spans l terminals if h is not a terminal, and $l + 1$ terminals otherwise.

For a terminal s, define the distance between s and h as the minimum weight of paths connecting s and h, where the weight of a path is the sum of the weights over all nodes on the path. Let $s_1, \ldots, s_l$ be the l terminals nearest to h with respect to the distance defined as above, and Q_i be the minimum weight path connecting h and s_i for each $i \in [l]$. Then the subgraph S' induced by $\bigcup_{i=1}^l V(Q_i)$ is a connected subgraph that spans at least l terminals. Let us show that the density of S' is at most D. Without loss of generality, let $w(P_1) \leq w(P_2) \leq \cdots \leq w(P_l)$. The definition of $s_1, \ldots, s_l$ indicates that $w(Q_i) \leq w(P_i)$ holds for each $i \in [l]$, so we have $w(S') = w(h) + \sum_{i=1}^l (w(Q_i) - w(h)) \leq w(h) + \sum_{i=1}^l (w(P_i) - w(h)) = w(S)$. Moreover, since the number of terminals spanned by S' is at least that spanned by S, the density of S' is at most D.

If we know h and l, the subgraph S' can be computed in polynomial time. Since there are only $|V|$ candidates for h, and $|T|$ candidates for l, we apply the algorithm for fixed h and l $O(|V||T|)$ times, and output the subgraph of minimum density from the obtained subgraphs. This gives a polynomial-time algorithm to compute a subgraph of density at most D spanning at least two terminals.   □

The details of the algorithm are described in Algorithm 1.

Algorithm 1 NWTREE

Input: an undirected graph $G = (V, E)$, node weights $w\colon V \to \mathbb{R}_+$, and a set of terminals $T \subseteq V$, ($w(t) = 0$ for $t \in T$)
Output: $U \subseteq V$ that induces a Steiner tree
1: $U \longleftarrow \emptyset$
2: **while** $|T| \geq 2$ **do**
3:     $S \longleftarrow$ MINDENSITY(G, w, T)
4:     $U \longleftarrow U \cup S$
5:     Shrink S into a single terminal of weight 0, and update G, w, and T
6: **end while**
7: output U

Algorithm 2 MINDENSTIY

Input: an undirected graph $G = (V, E)$, node weights $w: V \to \mathbb{R}_+$, and a set of terminals $T \subseteq V$ with $|T| \geq 2$
Output: $S \subseteq V$
1: $S \longleftarrow \emptyset$ (the density of S is defined as ∞)
2: **for** $h \in V$ **do**
3:    **for** $l = 2, \ldots, |T|$ **do**
4:       $s_1, \ldots, s_l \longleftarrow l$ terminals nearest to h
5:       for each $i \in [l]$, $Q_i \longleftarrow$ a minimum weight path connecting h and s_i
6:       $S' \longleftarrow \bigcup_{i=1}^{l} V(Q_i)$
7:       **if** density of $S' \leq$ density of S **then**
8:          $S \longleftarrow S'$
9:       **end if**
10:   **end for**
11: **end for**
12: output S

Theorem 2 *Algorithm* NWTREE *is an* $O(\log |T|)$*-approximation algorithm for the node-weighted Steiner tree problem.*

Proof Let S_i be the subgraph computed by MINDENSITY(G, w, T) in the i-th iteration of NWTREE, and let l be the number of iterations of NWTREE. Hence NWTREE outputs $\bigcup_{i=1}^{l} S_i$. Moreover, let T_i denote the set of terminals at the beginning of the i-th iteration for each $i \in [l]$. Recall that $T_1 = T$.

Let $i \in [l]$. By Theorem 1 and Lemma 2, the density of S_i is at most $\text{OPT}/|T_i|$. In other words, $w(S_i) \leq |T_i(S_i)| \cdot \text{OPT}/|T_i|$. When S_i is shrunk, the number of terminals decreased by $|T_i(S_i)| - 1$. Hence $|T_i| - (|T_i(S_i)| - 1) = |T_{i+1}|$ holds, where we let $|T_{l+1}| = 1$ by an abuse of notation. Since $|T_i(S_i)| \geq 2$, we have $|T_i(S_i)| \leq 2(|T_i(S_i)| - 1)$. From these, we have

$$\sum_{i=1}^{l} w(S_i) \leq \text{OPT} \cdot \sum_{i=1}^{l} \frac{|T_i(S_i)|}{|T_i|}$$

$$\leq 2\text{OPT} \cdot \sum_{i=1}^{l} \frac{|T_i(S_i)| - 1}{|T_i|}$$

$$\leq 2\text{OPT} \cdot \sum_{i=1}^{l} \left(\frac{1}{|T_i|} + \frac{1}{|T_i| - 1} + \cdots + \frac{1}{|T_i| - (|T_i(S_i)| - 2)} \right)$$

$$= 2\text{OPT} \cdot \left(\frac{1}{|T|} + \frac{1}{|T| - 1} + \cdots + \frac{1}{1} \right)$$

$$= O(\log |T|) \cdot \text{OPT}.$$

$\square$

To conclude this section, let us summarize the role of the spiders in the algorithm. The notion of spiders is useful because spiders have the follow two properties.

- The minimum density of spiders can be bounded by the density of an optimal solution through the spider decomposition theorem.
- There exists a polynomial-time algorithm for computing a subgraph the density of which is at most the minimum density of spiders.

4   LP-Based Analysis of the Spider Covering Algorithm

In Sect. 3, we observed that the spider covering framework achieves an $O(\log |T|)$-approximation algorithm. A key in the analysis is the spider decomposition theorem to bound the minimum density of spiders with respect to the density of an optimal solution. In this section, we present an alternate proof given by Guha et al. [5] for the fact that the density of spiders can be bounded. Specifically, we observe that there exists a spider of density at most $LP/|T|$, where LP is the optimal value of some LP relaxation of the problem. Since $LP \leq OPT$, this is indeed a stronger result than Theorem 1.

This LP-based proof has numerous advantages over the original proof based on the spider decomposition theorem. First, the proof can be easily extended to other problems, for which extending the spider decomposition theorem is often involved. We see in the next section that the algorithm for the node-weighted Steiner tree can be extended to an SNDP with higher connectivity requirements. In the analysis of this extension, we have to bound the minimum density of spiders, as well as extend the definition of the spiders. This can be done much easier if we consider the LP-based proof. Second, due to the LP-based proof, the algorithm can be easily combined with other techniques based on LPs, which makes it possible to modify the algorithm to apply to other related problems. For example, Klein and Ravi's algorithm can be easily modified to an $O(\log |T|)$-approximation algorithm for the prize-collecting node-weighted Steiner tree problem by using a standard threshold rounding [10]. Guha et al. [5] presented the LP-based proof motivated by this advantage. They designed new algorithms for several related problems, and those algorithms use Klein and Ravi's algorithm as a subroutine. To bound the approximation ratios of these algorithms, they needed to bound the minimum density of spiders in terms of the optimal LP value.

Now let us present the LP-based proof. As in the previous section, we assume that $w(t) = 0$ for all terminals $t \in T$. Moreover, no two terminals are adjacent; if two terminals were adjacent, then they would form a spider of density 0.

We define an LP relaxation of the node-weighted Steiner tree problem. Let us define $\mathscr{X}$ as $\{X \subseteq V : X \cap T \neq \emptyset \neq T \setminus (X \cup \Gamma_E(X)), \Gamma_E(X) \cap T = \emptyset\}$. We prepare a variable $x(v)$ for each $v \in V \setminus T$ to indicate whether v is spanned by a Steiner tree (if $x(v) = 1$, then v is spanned by a Steiner tree, and $x(v) = 0$ otherwise). For each $X \in \mathscr{X}$, a Steiner tree includes at least one node from $\Gamma_E(X)$, and hence $\sum_{v \in \Gamma_E(X)} x(v) \geq 1$ is a valid inequality. Then, the LP relaxation is formulated as follows.

$$\text{minimize} \quad \sum_{v \in V \setminus T} w(v)x(v)$$
$$\text{subject to} \quad \sum_{v \in \Gamma_E(X)} x(v) \geq 1, \quad \forall X \in \mathscr{X}, \tag{1}$$
$$x(v) \geq 0, \qquad\qquad \forall v \in V.$$

In the rest of this section, LP denotes the optimal value of (1). Since (1) relaxes the node-weighted Steiner tree, LP $\leq$ OPT holds. We prove the following theorem.

Theorem 3 *There exists a spider with density less than or equal to* LP$/|T|$.

Theorem 3 can be proven by presenting an algorithm that constructs the required spider. We present a primal-dual algorithm for this purpose. The following is the dual of (1).

$$\text{maximize} \quad \sum_{X \in \mathscr{X}} y(X)$$
$$\text{subject to} \quad \sum_{X \in \mathscr{X} : v \in \Gamma_E(X)} y(X) \leq w(v), \quad \forall v \in V \setminus T, \tag{2}$$
$$y(X) \geq 0, \qquad\qquad \forall X \in \mathscr{X}.$$

The algorithm maintains a node set $S_t \subseteq V$ for each $t \in T$ and a solution y for (2). Initially, S_t is set to $\{t\}$ for each $t \in T$, and $y(X) = 0$ for each $X \in \mathscr{X}$. Then, the algorithm repeats increasing y and adding a node to one of the node sets $S_t, t \in T$. During this process, y is always feasible for (2). Moreover,

$$(S_t \cup \Gamma_E(S_t)) \cap S_{t'} = \emptyset \tag{3}$$

holds for any distinct $t, t' \in T$ before the termination of the process. Notice that this condition holds when $S_t, t \in T$ are initialized because no two terminals are adjacent.

Let us explain how y and $S_t, t \in T$, are updated. We increase $y(S_t)$ uniformly for all $t \in T$ until the first constraint of (2) holds with equality for some $v \in V \setminus T$. We have two cases: (i) there exists exactly one terminal t such that $v \in \Gamma_E(S_t)$; (ii) $v \in \Gamma_E(S_t)$ holds for more than one terminal t. In Case (i), we add v to S_t, and continue the process with the updated node sets. Observe that this update preserves the condition (3).

In case (ii), we let $T' = \{t \in T : v \in \Gamma_E(S_t)\}$, add v to S_t for each $t \in T'$, and proceed to the following post-process. For each $t \in T'$, the subgraph induced by S_t includes a path between t and v. Let P_t be a shortest path in the induced subgraph, where the length of a path is defined by the node weights. The union of $P_t, t \in T'$, is a spider with head v and feet T'. The algorithm outputs this spider. The details of the algorithm are described in Algorithm 3.

In the following proof of Theorem 3, we show that the density of the spider output by the algorithm is at most $\sum_{X \in \mathscr{X}} y(X)/|T|$, where y denotes the dual solution when the algorithm terminates. Since y is feasible for (2), $\sum_{X \in \mathscr{X}} y(X) \leq$ LP holds, which gives Theorem 3.

Proof (Theorem 3) Suppose that the algorithm has l iterations, and the i-th iteration increases each of $y(S_t), t \in T$, by ϵ_i. Since each iteration increases $|T|$ dual variables, $\sum_{X \in \mathscr{X}} y(X) = |T| \sum_{i=1}^{l} \epsilon_i$ holds when the algorithm terminates.

Algorithm 3 LPSPIDER

Input: an undirected graph $G = (V, E)$, node weights $w \colon V \to \mathbb{R}_+$, and a set of terminals $T \subseteq V$
($w(t) = 0$ for $t \in T$, and no two terminals are adjacent)
Output: $S \subseteq V$ that induces a spider
1: $S_t \longleftarrow \{t\}$ for each $t \in T$, $y(X) \longleftarrow 0$ for each $X \in \mathscr{X}$
2: for each $v \in V \setminus T$, $w'(v) \longleftarrow w(v) - \sum_{X \in \mathscr{X} : v \in \Gamma_E(X)} y(X)$ and $c(v) \longleftarrow |\{t \in T : v \in \Gamma_E(S_t)\}|$
3: $v \longleftarrow \arg\min_{v \in V \setminus T} w'(v)/c(v)$
4: $y(S_t) \longleftarrow y(S_t) + w'(v)/c(v)$ for each $t \in T$
5: if $c(v) = 1$ then
6: $t \longleftarrow$ the terminal with $v \in \Gamma_E(S_t)$
7: $S_t \longleftarrow S_t \cup \{v\}$
8: go to Step 2
9: else
10: $T' \longleftarrow \{t \in T : v \in \Gamma_E(S_t)\}$
11: $S_t \longleftarrow S_t \cup \{v\}$ for each $t \in T'$
12: $P_t \longleftarrow$ a shortest path between v and t in $G[S_t]$ for each $t \in T'$
13: output $\bigcup_{t \in T'} V(P_t)$
14: end if

Below we prove that $\sum_{v \in \bigcup_{t \in T'} V(P_t)} w(v) \leq |T'| \sum_{i=1}^{l} \epsilon_i$. Since the feet of the spider are terminals in T', the density of the spider output by Algorithm 3 is at most $\sum_{i=1}^{l} \epsilon_i = \sum_{X \in \mathscr{X}} y(X)/|T| \leq \mathrm{LP}/|T|$.

Let $t \in T'$. For each $i \in [l]$, let S_t^i denote S_t in the i-th iteration. Then $t \in S_t^1 \subseteq S_t^2 \subseteq \cdots \subseteq S_t^l$. Let $P_t = (h = u_t^0, u_t^1, \ldots, u_t^p = t)$ be a path between h and t defined recursively as follows: for $j \in [p]$, let i be the minimum index such that $u_t^{j-1} \in \Gamma_E(S_t^i)$; u_t^j is defined as a node in S_t^j that is adjacent to u_t^{j-1}. Since P_t is a shortest path between h and t, $\sum_{v \in V(P_t) \setminus \{h\}} w(v) \leq \sum_{j=1}^{p} w(u_t^j)$ holds.

For any $t \in T'$ and $i \in [l]$, $\Gamma_E(S_t^i)$ includes exactly one node in $\bigcup_{t' \in T'} V(P_{t'})$. Moreover, any $v \in \bigcup_{t' \in T'} V(P_{t'})$ satisfies $\sum_{t' \in T', i \in [l] : v \in \Gamma_E(S_{t'}^i)} y(S_{t'}^i) = w(v)$. Therefore, $\sum_{v \in \bigcup_{t \in T'} V(P_t)} w(v) \leq \sum_{v \in \bigcup_{t \in T'} V(P_t)} w(v) \leq \sum_{t \in T'} \sum_{i \in [l]} y(S_t^i) = |T'| \sum_{i=1}^{l} \epsilon_i$. $\qquad\square$

5 Node-Weighted Edge-Connectivity Network Design

In this section, we consider an SNDP, which is an extension of the Steiner tree problem to a problem of constructing a network that satisfies higher connectivity requirements. As the connectivity requirements, we consider requirements on the edge-connectivity here.

Formally, the problem discussed in this section is defined as follows. Given an undirected graph G on a node set V, node weights $w \colon V \to \mathbb{R}_+$, and connectivity demands $r \colon \binom{V}{2} \to \mathbb{Z}_+$, we are asked to compute a minimum weight subset U of G such that the edge-connectivity of nodes $u, v \in V$ in $G[U]$ is at least $r(u, v)$ (i.e., there exist $r(u, v)$ edge-disjoint paths between u and v) for each u and v with $r(u, v) > 0$. This problem is called the edge-connectivity SNDP. Let $r_{\max} = \max_{u, v \in V} r(u, v)$ and $n = |V|$. We present an $O(r_{\max} \log n)$-approximation algorithm for the edge-connectivity SNDP, which was developed by Nutov [6].

A popular approach for solving an SNDP is to reduce the problem to a special case called the augmentation problem. In the augmentation problem, we are given a subset V_0 of V and node pairs $D \subseteq \binom{V_0}{2}$ such that $\lambda_{G[V_0]}(u, v) = k - 1$ for each $\{u, v\} \in D$. The objective is to find a minimum weight node set $U \subseteq V \setminus V_0$ such that $\lambda_{G[V_0 \cup U]}(u, v) \geq k$ for each $\{u, v\} \in D$. This is the special case of the edge-connectivity SNDP in which $w(v) = 0$ for all $v \in V_0$. We can solve the SNDP by repeatedly solving the augmentation problem, setting k from 1 through $r_{\max}$. Specifically, if an α-approximation algorithm for the augmentation problem is available, then we have an $\alpha r_{\max}$-approximation algorithm for the edge-connectivity SNDP.

In the rest of this section, we present an $O(\log n)$-approximation algorithm for the augmentation problem. We assume without loss of generality that $w(v) = 0$ for all $v \in V_0$. Let E_0 be the set of edges induced by V_0, and E be the set of the other edges in G. If a subset X of V satisfies $|X \cap \{u, v\}| = |\{u, v\} \setminus X| = 1$ for $\{u, v\} \in D$, then we say that X separates $\{u, v\}$. Since $\lambda_{G[V_0]}(u, v) = k - 1$, all $X \subset V$ separating $\{u, v\} \in D$ satisfy $|\delta_{E_0}(X)| \geq k - 1$. We let $\mathscr{X}_{u,v}$ denote $\{X \subset V : |X \cap \{u, v\}| = |\{u, v\} \setminus X| = 1, |\delta_{E_0}(X)| = k - 1\}$ for each $\{u, v\} \in D$, and denote $\bigcup_{\{u,v\} \in D} \mathscr{X}_{u,v}$ by $\mathscr{X}$. A node set $U \subseteq V \setminus V_0$ is feasible for the augmentation problem if $V_0 \cup U$ induces an edge set $F \subseteq E$ such that $\delta_F(X) \neq \emptyset$ for each $X \in \mathscr{X}$. Hence the augmentation problem is equivalent to computing an edge set $F \subseteq E$ such that $\delta_F(X) \neq \emptyset$ for each $X \in \mathscr{X}$, and $\sum_{v \in V(F)} w(v)$ is minimized. We say that an edge set F covers a family $\mathscr{X}$ if $\delta_F(X) \neq \emptyset$ for each $X \in \mathscr{X}$.

A family $\mathscr{V} \subseteq 2^V$ is called *uncrossable* if, for any $X, Y \in \mathscr{V}$, we have $X \cap Y, X \cup Y \in \mathscr{V}$ or $X \setminus Y, Y \setminus X \in \mathscr{V}$.

Lemma 3 *$\mathscr{X}$ is uncrossable.*

Proof Let $X, Y \in \mathscr{X}$. The submodularity and posimodularity of cut functions of undirected graphs indicate that the following two inequalities holds.

$$|\delta_{E_0}(X)| + |\delta_{E_0}(Y)| \geq |\delta_{E_0}(X \cap Y)| + |\delta_{E_0}(X \cup Y)| \tag{4}$$

and

$$|\delta_{E_0}(X)| + |\delta_{E_0}(Y)| \geq |\delta_{E_0}(X \setminus Y)| + |\delta_{E_0}(X \setminus Y)| \tag{5}$$

By the definition of $\mathscr{X}$, the left-hand sides of these inequalities are equal to $2(k - 1)$. Moreover, by a case analysis, we can observe that (i) both $X \cap Y$ and $X \cup Y$ separate at least one demand pair, or (ii) both $X \setminus Y$ and $Y \setminus X$ do. In the former case, the right-hand side of (4) is at least $2(k - 1)$. Hence the inequality (4) holds with equality, implying that $|\delta_{E_0}(X \cap Y)| = |\delta_{E_0}(X \cup Y)| = k - 1$. This means that $X \cap Y, X \cup Y \in \mathscr{X}$. In the latter case, the right-hand side of (5) is at least $2(k - 1)$. Similar to the former case, this indicates that $X \setminus Y, Y \setminus X \in \mathscr{X}$. □

By Lemma 3, the augmentation problem is a special case of the following problem: Given a finite set V, a set E of undirected edges on V, node weights $w \colon V \to \mathbb{R}_+$, and an uncrossable family $\mathcal{V} \subseteq 2^V$, find a subset F of E that covers $\mathcal{V}$, and minimizes the weight $w(F)$. We call this problem the *uncrossable family covering problem*. In the rest of this section, we refer to an arbitrary uncrossable family as $\mathcal{V}$, and to a general uncrossable family arising from the augmentation problem as $\mathcal{X}$.

$X \in \mathcal{V}$ is called a *min-core* of $\mathcal{V}$ if it is inclusion-wise minimal in $\mathcal{V}$, and is called a *core* if it includes only one min-core as a subset. Note that each min-core is also a core. For a family $\mathcal{C}$ of cores and a min-core X, we let $\mathcal{C}(X)$ denote $\{Y \in \mathcal{C} \colon X \subseteq Y\}$. The min-cores and cores of an uncrossable family $\mathcal{V}$ constitute certain nice structures. The following facts are examples of useful conditions derived from the structure.

- Let X be a min-core of $\mathcal{V}$. Then $\mathcal{C}(X)$ is a ring family; i.e., any $X', X'' \in \mathcal{C}(X)$ satisfy $X' \cap X'', X' \cup X'' \in \mathcal{C}(X)$.
- Let X and Y be distinct min-cores of $\mathcal{V}$. If $X' \in \mathcal{C}(X)$, then $X' \cap Y = \emptyset$ holds. In particular, min-cores of $\mathcal{V}$ are pairwise disjoint.

Moreover, if $\mathcal{V}$ is an uncrossable family and F is an edge set, then the residual family $\mathcal{V}_F := \{X \in \mathcal{V} \colon \delta_F(X) = \emptyset\}$ is also uncrossable. These properties are used implicitly in the following discussion.

For a node h and a min-core X in a core family $\mathcal{C}$, let $\mathcal{C}(X, h) = \{Y \in \mathcal{C}(X) \colon h \notin Y\}$. Now, we define a spider used in the $O(\log n)$-approximation algorithm for the uncrossable family covering problem. Nutov [6] extended the notion of spiders introduced by Klein and Ravi [2] as follows.

Definition 2 For an uncrossable family $\mathcal{V}$ with a core family $\mathcal{C}$, a spider is an edge set S for which there exists a node $h \in V(S)$ and min-cores $X_1, \ldots, X_f$ such that

- F can be decomposed into disjoint subsets $F_1, \ldots, F_f$,
- F_i covers $\mathcal{C}(X_i, h)$ for each $i \in [f]$,
- $V(S_i) \cap V(S_j) \subseteq \{h\}$ holds for $i, j \in [f]$ with $i \neq j$,
- and $\mathcal{C}(X_1, h) = \mathcal{C}(X_1)$ when $f = 1$.

Here, h is called the head of S, and $X_1, \ldots, X_f$ are called the feet of S. The density of S is defined as $w(S)/f$.

Let OPT denote the minimum weight of feasible solutions for the augmentation problem. Nutov [6] proved that any cover F of an uncrossable family $\mathcal{V}$ can be decomposed into pairwise-disjoint spiders so that each min-core is a foot of some spider. This means that there exists a spider of density at most $\text{OPT}/|\mathcal{C}|$. Here, we present an alternative LP-based proof for this fact given by Chekuri, Ene, and Vakilian [11]. This proof uses the following property of uncrossable families arising from the augmentation problem: if $Y \subseteq V \setminus V_0$ and X is a minimal member of $\mathcal{X}$ such that $Y \cap \Gamma_E(X) = \emptyset$, then $X \subseteq Y \cup V_0$. Hence their proof cannot be applied to arbitrary uncrossable families, but it can be applied to those arising from the augmentation problem. Hereinafter, $\mathcal{C}$ and $\mathcal{M}$ denote the families of cores and min-cores of the uncrossable family $\mathcal{X}$ arising from the augmentation problem.

For $X \subset V$, let $\Gamma_E(X)$ denote the set of nodes in $V \setminus X$ to which some edges in $\delta_E(X)$ are incident. We use the following LP to bound the minimum density of spiders.

$$
\begin{aligned}
\text{minimize} \quad & \sum_{v \in V} w(v)x(v) \\
\text{subject to} \quad & \sum_{v \in \Gamma_E(X)} x(v) \geq 1, \quad \forall X \in \mathscr{X}, \\
& x(v) \geq 0, \qquad\qquad \forall v \in V.
\end{aligned}
\tag{6}
$$

Actually, (6) is an LP relaxation of the augmentation problem. To see this, from a feasible solution $F \subseteq E$ for the augmentation problem, define a solution $x: V \to \{0, 1\}$ so that $x(v) = 1$ if and only if $v \in V(F)$. Then x is feasible for (6), and $\sum_{v \in V} w(v)x(v) = w(F)$ holds. We let LP denote the optimal value of (6). The following theorem indicates that the minimum density of spiders is at most $\text{OPT}/|\mathscr{C}|$ because $\text{LP} \leq \text{OPT}$.

Theorem 4 *There exists a spider S of density at most $\text{LP}/|\mathscr{M}|$.*

Theorem 4 is proven by presenting a primal-dual algorithm to construct the required spider.

The primal-dual algorithm is similar to Algorithm (3); therefore, we do not present the details of the algorithm; refer to Chekuri et al. [11] for the proof of Theorem 4.

Let us see how Theorem 4 provides an $O(\log n)$-approximation algorithm for the augmentation problem.

For an edge set S, we define a number f_S as follows. If there exists a node h such that S covers $\mathscr{C}(X, h)$ for at least two min-cores X, f_S is the maximum number of such min-cores, where h is taken over all nodes. If there exists no such node h, and if S covers $\mathscr{C}(X)$ for some min-core X, then $f_S = 1$. Otherwise, $f_S = 0$. We define the density of S as $w(S)/f_S$.

To provide an $O(\log n)$-approximation algorithm for the augmentation problem, we prove two preparatory lemmas.

Lemma 4 *There exists a polynomial-time algorithm to compute an edge set S whose density is at most $O(1) \cdot \text{OPT}/|\mathscr{M}|$.*

Proof The primal-dual algorithm used for proving Theorem 4 satisfies the required condition. Simultaneously, we have an alternative algorithm, which is defined as follows. Suppose that the head of a minimum density spider S^* is h, and the number of feet of S^* is f. Since h and f can be guessed in $O(n|\mathscr{C}|)$ time, we can assume that we know h and f. We consider the problem of computing a minimum weight edge set covering $\mathscr{C}(X, h)$ for a min-core X. Note that $\mathscr{C}(X, h)$ is a ring family. Using a property of ring families, we can reduce this covering problem to the problem with edge weights with a loss of factor 2, and the covering problem with edge weights admits an exact polynomial-time algorithm [6]. Hence there exists a polynomial-time algorithm to compute an edge set S_X covering $\mathscr{C}(X, h)$ such that $w(S_X)$ is at most twice the minimum weight of covers of $\mathscr{C}(X, h)$. We compute S_X for each $X \in \mathscr{C}$, and then choose f min-cores $X_1, \ldots, X_f$ for which the weights $w(S_{X_i})$, $i \in [f]$, are smaller than the others. In this process, we set the weight of h

to 0. Then we can observe that the density of an edge set $S := \bigcup_{i=1}^{f} S_{X_i}$ is at most $2\text{OPT}/|\mathcal{M}|$. Indeed, the spider S^* can be decomposed into $S_1^*, \ldots, S_f^*$ so that each S_i^* is a covering of $\mathcal{C}(X, h)$ for some min-core X, and $V(S_i^*) \cap V(S_j^*) \subseteq \{h\}$ for $i \neq j$. Hence, $w(S) \leq w(h) + \sum_{i=1}^{f} w(V(S_{X_i}) \setminus \{h\}) \leq w(h) + 2\sum_{i=1}^{f} w(V(S_i^*) \setminus \{h\}) \leq 2w(S^*)$. This indicates that the density of S is at most twice the density of S^*. By Theorem 4, the density of S^* is at most $\text{OPT}/|\mathcal{M}|$.   □

Lemma 5 *Let S be an edge set S with $f_S \geq 1$, and let $\mathcal{M}_S$ denote the min-core family of $\{X \in \mathcal{X} : \delta_S(X) = \emptyset\}$. Then, $|\mathcal{M}| - |\mathcal{M}_S| \geq \max\{1, \lceil (f_S - 1)/2 \rceil\} \geq f_S/3$.*

Proof Each min-core in $\mathcal{M}_S$ includes at least one min-core in $\mathcal{M}$. Let t be the number of min-cores in $\mathcal{M}_S$ that include exactly one min-core in $\mathcal{M}$. Recall that min-cores in $\mathcal{M}_S$ are pairwise disjoint. Hence $|\mathcal{M}| - |\mathcal{M}_S| \geq \lceil (|\mathcal{M}| - t)/2 \rceil$. Let X be a min-core in $\mathcal{M}_S$ counted in t.

If $f_S = 1$, then S covers $\mathcal{C}(Y)$ for some min-core $Y \in \mathcal{M}$, and hence $X \in \mathcal{C}(Y')$ holds for some $Y' \in \mathcal{M} \setminus \{Y\}$. Note that at most one min-core in $\mathcal{M}_S$ belongs to $\mathcal{C}(Y')$ for each $Y' \in \mathcal{M} \setminus \{Y\}$. Hence $t \leq |\mathcal{M}| - 1$.

Let $f_S \geq 2$. Let $Y \in \mathcal{M}$ such that $\mathcal{C}(Y, h)$ is covered by S. If $X \in \mathcal{C}(Y)$, then $h \in X$ because X is not covered by S. Since min-cores in $\mathcal{M}_S$ are pairwise disjoint, at most one min-core in $\mathcal{M}_S$ includes h. Thus, $t \leq |\mathcal{M}| - f_S + 1$.   □

Theorem 5 *The augmentation admits an $O(\log |\mathcal{M}|)$-approximation algorithm.*

Proof We show that $O(\log |\mathcal{M}|)$-approximation is achieved by the algorithm of choosing edge sets computed by the algorithm in Lemma 4 repeatedly until all sets in $\mathcal{X}$ are covered.

Suppose that l is the number of iterations in the algorithm. For each $i \in [l]$, let S_i be the edge set computed in the i-th iteration, and let $\mathcal{M}_i$ be the family of min-cores at the beginning of the i-th iteration. Then, $w(S_i)/f_{S_i} \leq \alpha \cdot \text{OPT}/|\mathcal{M}_i|$ and $|\mathcal{M}_i| - |\mathcal{M}_{i+1}| \geq f_{S_i}/3$ hold for each $i \in [l]$ and some constant α by Lemmas 4 and 5, where we let $|\mathcal{M}_{l+1}| = 0$ for convenience. Hence,

$$\sum_{i=1}^{l} w(S_i) \leq \alpha \cdot \text{OPT} \cdot \sum_{i=1}^{l} \frac{f_{S_i}}{|\mathcal{M}_i|}$$

$$\leq 3\alpha \cdot \text{OPT} \sum_{i=1}^{l} \frac{|\mathcal{M}_i| - |\mathcal{M}_{i+1}|}{|\mathcal{M}_i|}$$

$$\leq 3\alpha \cdot \text{OPT} \sum_{i=1}^{|\mathcal{M}_1|} \frac{1}{i}$$

$$= O(\log |\mathcal{M}|) \cdot \text{OPT}.$$

□

6 Node-Connectivity Network Activation Problem

In this section, we summarize the previous work of the author on the node-connectivity network activation problem [12].

The *network activation problem* is defined as follows. We are given an undirected graph $G = (V, E)$ and a set W of non-negative real numbers such that $0 \in W$. A solution in the problem is a node weight function $w : V \to W$. Each edge $\{u, v\} \in E$ is associated with an activation function $\psi^{uv} : W \times W \to \{\text{true, false}\}$ such that $\psi^{uv}(i, j) = \psi^{vu}(j, i)$ holds for any $i, j \in W$. Here, each activation function ψ^{uv} is supposed to be *monotone*; i.e., if $\psi^{uv}(i, j) = \text{true}$ for some $i, j \in W$, then $\psi^{uv}(i', j') = \text{true}$ for any $i', j' \in W$ with $i' \geq i$ and $j' \geq j$. An edge $\{u, v\}$ is *activated* by w if $\psi^{uv}(w(u), w(v)) = \text{true}$. Let E_w be the set of edges activated by w in E. A node weight function w is feasible in the network activation problem if E_w satisfies given constraints, and the objective of the problem is to find a feasible node weight function w that minimizes $w(V)$.

In this section, we consider the node-connectivity constraints on the set E_w of activated edges. Namely, we are given connectivity demands $r : \binom{V}{2} \to \mathbb{Z}_+$, and the node-connectivity between two nodes u and v must be at least $r(u, v)$ (i.e., there exist $r(u, v)$ inner disjoint paths between u and v) in the graph (V, E_w) for every $u, v \in V$. We denote $\max_{u,v \in V} r(u, v)$ by $r_{\max}$ and $|V|$ by n. If we are given node weights $w' : V \to \mathbb{Z}_+$, W is defined as $\{w'(v) : v \in V\}$, and each edge uv is associated with an activation function ψ^{uv} defined so that $\psi^{uv}(i, j) = \text{true}$ if $i \geq w'(u)$ and $j \geq w'(v)$, then the network activation problem is equivalent to the node-weighted SNDP with node-connectivity constraints. Hence the network activation problem is an extension of an SNDP.

The network activation problem was introduced by Panigrahi [13]. He gave $O(\log n)$-approximation algorithms for $r_{\max} \leq 2$, and proved that it is NP-hard to obtain an $o(\log n)$-approximation algorithm even when activated edges are required to constitute a spanning tree. Nutov [8] presented an $O(r_{\max}^4 \log^2 n)$-approximation algorithm for general $r_{\max}$. This result is based on his research on the node-weighted SNDP in [7]. However, this result in [7]. contains an error. A correction is provided by Vakilian [14] for the node-weighted SNDP, but it cannot be extended to the network activation problem. In this section, we present an $O(r_{\max}^5 \log^2 n)$-approximation algorithm given by the author in [12]. Actually, the author presented there an algorithm for the more general prize-collecting network activation problem, but we do not discuss that direction in the present article.

To solve the network activation problem, we reduce it to the augmentation problem as we did in the previous section for the node-weighted SNDP. Here, the augmentation problem is defined as follows. We are given two edge sets E_0 and E, and activation functions are given for edges in E. We are also given a demand set $D \subseteq \binom{V}{2}$ such that the node-connectivity of each demand pair $\{s, t\} \in D$ is at least $k - 1$ in the graph (V, E_0). Then a subset F of E is feasible if the connectivity of each demand pair in $(V, E_0 \cup F)$ is at least k. The objective of the problem is to find a node weight function $w : V \to W$ so that E_w is feasible and $w(V)$ is minimized. As in SNDPs, if an α-approximation algorithm is available for the augmentation problem, we have an $\alpha r_{\max}$-approximate solution for the the the network activation problem.

The augmentation problem can be defined as a problem of activating edges covering bisets, where a *biset* is an ordered pair $\hat{X} = (X, X^+)$ of subsets of V such that $X \subseteq X^+$. The former element of a biset is called the *inner-part* and the latter is called the *outer-part*. We always let X denote the inner-part of a biset $\hat{X}$ and X^+ denote the outer-part of $\hat{X}$. $X^+ \setminus X$ is called the *boundary* of a biset $\hat{X}$ and is denoted by $\Gamma(\hat{X})$. For an edge set E, $\delta_E(\hat{X})$ denotes the set of edges in E that have one end-node in X and the other in $V \setminus X^+$. We say that an edge e *covers* $\hat{X}$ if $e \in \delta_E(\hat{X})$, and a set F of edges *covers* a biset family $\mathcal{V}$ if each $\hat{X} \in \mathcal{V}$ is covered by some edge in F.

Let $i \in [d]$. We say that a biset $\hat{X}$ *separates* a demand pair $\{s_i, t_i\}$ if $|X \cap \{s_i, t_i\}| = |\{s_i, t_i\} \setminus X^+| = 1$. We define $\mathcal{V}_i$ as the family of bisets $\hat{X}$ such that $|\delta_{E_0}(\hat{X})| + |\Gamma(\hat{X})| = k - 1$ and $\hat{X}$ separates the demand pair $\{s_i, t_i\}$. $F \subseteq E$ increases the node-connectivity of $\{s_i, t_i\}$ from $k - 1$ to at least k if and only if F covers $\mathcal{V}_i$ (and this is equivalent to covering all bisets $\hat{X}$ with $\delta_{E_0}(\hat{X}) = \emptyset$ and $|\Gamma(\hat{X})| = k - 1$). Hence the augmentation problem is equivalent to the problem of activating edges covering the biset family $\mathcal{V} := \bigcup_{i \in [d]} \mathcal{V}_i$. We call this problem the *biset covering problem*.

For two bisets $\hat{X}$ and $\hat{Y}$, we define $\hat{X} \cap \hat{Y} = (X \cap Y, X^+ \cap Y^+)$, $\hat{X} \cup \hat{Y} = (X \cup Y, X^+ \cup Y^+)$, and $\hat{X} \setminus \hat{Y} = (X \setminus Y^+, X^+ \setminus Y)$. A biset family $\mathcal{V}$ is called *uncrossable* if, for any $\hat{X}, \hat{Y} \in \mathcal{V}$, (i) $\hat{X} \cap \hat{Y}, \hat{X} \cup \hat{Y} \in \mathcal{V}$, or (ii) $\hat{X} \setminus \hat{Y}, \hat{Y} \setminus \hat{X} \in \mathcal{V}$ holds. A biset family $\mathcal{V}$ is called a *ring-family* if (i) holds for any $\hat{X}, \hat{Y} \in \mathcal{V}$. A maximal biset in a ring-family is unique because ring-families are closed under union.

The biset family defined from the augmentation problem is not necessarily uncrossable. However, it is known that the family can be decomposed into $O(k^3 \log |V|)$ uncrossable families [15]. Hence, an α-approximation algorithm for the biset covering problem with uncrossable biset families gives an $O(\alpha k^3 \log |V|)$-approximation algorithm for the augmentation problem with the node-connectivity constraints. If the demand pairs satisfy certain conditions, this approximation ratio can be improved [7, 16]. Therefore, we consider the biset covering problem with uncrossable biset families in the rest of this section.

Our results require several properties of uncrossable biset families. We say that bisets $\hat{X}$ and $\hat{Y}$ are *strongly disjoint* when both $X \cap Y^+ = \emptyset$ and $X^+ \cap Y = \emptyset$ hold. When $X \subseteq Y$ and $X^+ \subseteq Y^+$, we say $\hat{X} \subseteq \hat{Y}$. Minimality and maximality in a biset family are defined with regard to inclusion. A biset family $\mathcal{V}$ is called *strongly laminar* when, if $\hat{X}, \hat{Y} \in \mathcal{V}$ are not strongly disjoint, then they are comparable (i.e., $\hat{X} \subseteq \hat{Y}$ or $\hat{Y} \subseteq \hat{X}$). A minimal biset in a biset family $\mathcal{V}$ is called a *min-core*, and $\mathcal{M}_{\mathcal{V}}$ denotes the family of min-cores in $\mathcal{V}$. A biset is called a *core* if it includes only one min-core, and $\mathcal{C}_{\mathcal{V}}$ denotes the family of cores in $\mathcal{V}$, where min-cores are also cores. When $\mathcal{V}$ is clear from the context, we may simply denote them by $\mathcal{M}$ and $\mathcal{C}$.

For a biset family $\mathcal{V}$, biset $\hat{X}$, and node v, $\mathcal{V}(\hat{X})$ denotes $\{\hat{Y} \in \mathcal{V} : \hat{X} \subseteq \hat{Y}\}$ and $\mathcal{V}(\hat{X}, v)$ denotes $\{\hat{Y} \in \mathcal{V}(\hat{X}) : v \notin Y^+\}$.

We define a spider for the biset family covering problem as follows.

Definition 3 A spider for a biset family $\mathcal{V}$ is an edge set $S \subseteq E$ such that there exist $h \in V$ and $\hat{X}_1, \ldots, \hat{X}_f \in \mathcal{M}$, and S can be decomposed into subsets $S_1, \ldots, S_f$ that satisfy the following conditions:

- $V(S_i) \cap V(S_j) \subseteq \{h\}$ for each $i, j \in [f]$ with $i \neq j$;
- S_i covers $\mathcal{C}(\hat{X}_i, h)$ for each $i \in [f]$;
- if $f = 1$, then $\mathcal{C}(\hat{X}_1, h) = \mathcal{C}(\hat{X}_1)$;
- $h \in X_i^+$ holds for at most one $i \in [f]$.

h is called the head, and $\hat{X}_1, \ldots, \hat{X}_f$ are called the feet of the spider. For a spider S, we let $f(S)$ denote the number of its feet.

This definition of spiders for biset families is given by the author in [12]. Before [12], a spider was considered by Nutov [7] for the biset family covering problem, but the definition in [12] is slightly different from the original one. In [7], an edge set is a spider even if it does not satisfy the last condition given above. However, the analysis on the spider covering algorithm with the original definition in [7] has an error. Hence we need to use the definition in [12].

What we have to do is to bound the minimum density. We present an LP-based analysis for this. First, let us describe the LP relaxation of the problem that we use. In this relaxation, we have a variable $x(uv, j, j')$ for each $uv \in A$ and $(j, j') \in \Psi^{uv}$. $x(uv, j, j') = 1$ indicates that an edge $\{u, v\}$ is activated by assigning weight j to node u and weight j' to node v. In addition, we have two variables $x_{in}(v, j)$ and $x_{out}(v, j)$ for each pair of $v \in V$ and $j \in W$, where $x_{in}(v, j) = 1$ indicates that v is assigned the weight j for activating edges entering v, and $x_{out}(v, j) = 1$ indicates that v is assigned the weight j for activating edges leaving v. Then the LP that we use is described as follows.

$$\text{minimize} \quad \sum_{v \in V} \sum_{j \in W} j \cdot (x_{in}(v, j) + x_{out}(v, j))$$

$$\text{subject to} \quad \sum_{uv \in \delta_A^-(\hat{X})} \sum_{(j,j') \in \Psi^{uv}} x(uv, j, j') \geq 1 \qquad \text{for} \hat{X} \in \mathcal{V}, \qquad (7)$$

$$x_{out}(u, j) \geq \sum_{\substack{v \in X: \\ uv \in A}} \sum_{\substack{j' \in W: \\ (j,j') \in \Psi^{uv}}} x(uv, j, j') \qquad \text{for} \hat{X} \in \mathcal{V}, u \in V \setminus X^+, j \in W,$$

$$\qquad (8)$$

$$x_{in}(v, j') \geq \sum_{\substack{u \in V \setminus X^+: \\ uv \in A}} \sum_{\substack{j \in W: \\ (j,j') \in \Psi^{uv}}} x(uv, j, j') \qquad \text{for} \hat{X} \in \mathcal{V}, v \in X, j' \in W, \quad (9)$$

$$x_{in}(v, j) \geq 0 \qquad \qquad \text{for } v \in V, j \in W,$$

$$x_{out}(v, j) \geq 0 \qquad \qquad \text{for } v \in V, j \in W,$$

$$x(uv, j, j') \geq 0 \qquad \qquad \text{for } uv \in A, (j, j') \in \Psi^{uv}.$$

Let $LP(\mathcal{V})$ denote the optimal objective value of this LP. The constraint (7) demands that each $\hat{X} \in \mathcal{V}$ be covered by at least one activated edge. If a solution x is required to satisfy (7), and conditions $x_{out}(u, j) \geq x(uv, j, j')$ and $x_{in}(v, j') \geq x(uv, j, j')$

for each $uv \in A$ and $(j, j') \in \Psi^{uv}$, then the objective function of the LP is at most
the optimal objective value of the biset covering problem. However, instead of these
conditions, the LP includes (8) and (9) as constraints. These constraints are stronger
than those required for the biset covering problem. Indeed, LP($\mathcal{V}$) does not relax
the biset covering problem. In fact, we do not use LP($\mathcal{V}$), but LP($\mathcal{L}$) defined from
some subfamily $\mathcal{L}$ of the core family $\mathcal{C}$. We do not know $\mathcal{L}$ beforehand, but we can
show that $\mathcal{L}$ is a strongly laminar subfamily of $\mathcal{C}$. The following lemma indicates
that in this case LP($\mathcal{L}$) is within a constant factor of OPT.

Lemma 6 LP($\mathcal{L}$) $\leq$ 2OPT *if $\mathcal{L}$ is a strongly laminar family of cores of $\mathcal{V}$.*

The dual of the LP is

$$\text{maximize} \quad \sum_{\hat{X} \in \mathcal{V}} z(\hat{X})$$

$$\text{subject to} \quad \sum_{\hat{X} \in \mathcal{V}:uv \in \delta_A^-(\hat{X})} z(\hat{X}) \leq \sum_{\hat{X} \in \mathcal{V}:uv \in \delta_A^-(\hat{X})} \left(z(\hat{X}, u, j) + z(\hat{X}, v, j') \right)$$

$$\text{for } uv \in A, (j, j') \in \Psi^{uv},$$

$$\sum_{\hat{X} \in \mathcal{V}:v \in X} z(\hat{X}, v, j') \leq j' \qquad \text{for } v \in V, j' \in W,$$

$$\sum_{\hat{X} \in \mathcal{V}:u \in V \setminus X^+} z(\hat{X}, u, j) \leq j \qquad \text{for } u \in V, j \in W,$$

$$z(\hat{X}) \geq 0 \qquad \text{for } \hat{X} \in \mathcal{V},$$

$$z(\hat{X}, v, j) \geq 0 \qquad \text{for } \hat{X} \in \mathcal{V}, v \notin \Gamma(\hat{X}), j \in W.$$

We can design a primal-dual algorithm to compute a weight function $w: V \to W$,
a spider S activated by w, and a solution z for the dual that satisfy the following
conditions:

- $w(V)/f(S) \leq \sum_{\hat{X} \in \mathcal{V}} z(\hat{X})/|\mathcal{M}|$,
- there exists a strongly laminar family $\mathcal{L}$ of cores such that $z(\hat{X}) > 0$ only for
  $\hat{X} \in \mathcal{L}$.

Such a solution z is also feasible to the dual of the LP obtained by replacing $\mathcal{V}$ with
$\mathcal{L}$. Hence $\sum_{\hat{X} \in \mathcal{V}} z(\hat{X}) = \sum_{\hat{X} \in \mathcal{L}} z(\hat{X}) \leq$ LP($\mathcal{L}$) holds, and the first condition and
Lemma 6 indicates that $w(V)/f(S) \leq 2$OPT$/|\mathcal{M}|$ holds. Therefore, we have the
following theorem.

Theorem 6 *Let $\mathcal{V}$ be an uncrossable family of bisets. There exists a polynomial-
time algorithm for computing $w: V \to W$ such that E_w contains a spider S and
$w(V)/f(S) \leq 2$OPT$/|\mathcal{M}_{\mathcal{V}}|$ holds.*

We do not present the primal-dual algorithm in this article. Refer to [12] for the
details of the algorithm.

The approximation algorithm for the biset covering problem basically repeats computing node weight functions guaranteed in Theorem 6, and outputs the component-wise maximum of the functions. This output function activates all the spiders chosen in the iterations because the activation functions are monotone.

For analyzing the performance of this algorithm, we need a potential function that measures the progress of the algorithm. For analyzing previous algorithms, the number of min-cores was used as the potential function. Even for the biset covering problem, if we proved that $|\mathcal{M}_\mathcal{V}| - |\mathcal{M}_{\mathcal{V}_S}| \geq f(S)/O(1)$ holds for each spider S of $\mathcal{V}$, we would have an $O(\log d)$-approximation algorithm. However, there is a case with $|\mathcal{M}_\mathcal{V}| - |\mathcal{M}_{\mathcal{V}_S}| = 0$ as follows. Let $\mathcal{V} = \{\hat{X}_1, \hat{Y}_1, \ldots, \hat{X}_n, \hat{Y}_n\}$, and suppose that $\hat{X}_l \subseteq \hat{Y}_l$ for each $l \in [n]$, $\hat{Y}_l$ and $\hat{Y}_{l'}$ are strongly disjoint for each $l, l' \in [n]$ with $l \neq l'$, and a node h is in $\Gamma(\hat{Y}_l) \setminus X_l^+$ for each $l \in [n]$. $\mathcal{V}$ is strongly laminar, and hence uncrossable. Note that $\mathcal{M}_\mathcal{V} = \{\hat{X}_1, \ldots, \hat{X}_n\}$, and hence $|\mathcal{M}_\mathcal{V}| = n$. If the head of a spider S is h and its feet are $\hat{X}_1, \ldots, \hat{X}_n$ (i.e., $f(S) = n$), then $\mathcal{M}_{\mathcal{V}_S} = \{\hat{Y}_1, \ldots, \hat{Y}_n\}$ holds, and hence $|\mathcal{M}_{\mathcal{V}_S}| = n$. Therefore, $|\mathcal{M}_\mathcal{V}| - |\mathcal{M}_{\mathcal{V}_S}| = 0$.

Vakilian [14] showed that such an inconvenient situation does not appear if $\mathcal{V}$ arises from the node-weighted SNDP. We have already seen this in Sect. 5 when the connectivity constraints are defined with regard to the edge-connectivity. To explain that it is true for an SNDP with node-connectivity constraints, let (V, E_0) be the graph to be augmented in an instance of the augmentation problem. Recall that the problem requires adding edges in an edge set E to E_0. If this instance is obtained by the reduction from the node-weighted SNDP, then E_0 is the subset of $E_0 \cup E$ induced by some node set $U \subseteq V$, and each biset $\hat{X}$ that is required to be covered satisfies $\Gamma(\hat{X}) \subseteq U$. Moreover, a spider is not chosen if its head is in U, and therefore the heads of chosen spiders are not included in the boundary of any biset. This means that each spider S achieves $|\mathcal{M}_\mathcal{V}| - |\mathcal{M}_{\mathcal{V}_S}| \geq f(S)/3$ for $\mathcal{V}$ arising from the node-weighted SNDP. However, this is not the case for all uncrossable biset families, including those arising from the network activation problem because (V, E_0) may not be an induced subgraph in general.

Therefore, using the number of min-cores as a potential function gives no desired approximation guarantee for general uncrossable biset families. For this reason, the author introduced a new potential function in [12]. We introduce this potential function in the following.

Let γ denote $\max_{\hat{X} \in \mathcal{V}} |\Gamma(\hat{X})|$. If $\mathcal{V}$ arises from the augmentation problem, $\gamma \leq k - 1$ holds. For a family $\mathcal{X}$ of cores and core $\hat{X} \in \mathcal{X}$, let $\Delta_\mathcal{X}(\hat{X})$ denote the set of nodes $v \in \Gamma(\hat{X})$ such that there exists another core $\hat{Y} \in \mathcal{X} \setminus \{\hat{X}\}$ with $v \in \Gamma(\hat{Y})$. We define the potential $\phi_\mathcal{X}(\hat{X})$ of a core $\hat{X}$ as $\gamma - |\Delta_\mathcal{X}(\hat{X})|$. The potential $\phi(\mathcal{X})$ of $\mathcal{X}$ is defined as $(\gamma + 1)|\mathcal{X}| + \sum_{\hat{X} \in \mathcal{X}} \phi_\mathcal{X}(\hat{X})$.

Lemma 7 ([12]) Let S be a spider for $\mathcal{V}$. If $f(S) = 1$, then $\phi(\mathcal{M}_\mathcal{V}) - \phi(\mathcal{M}_{\mathcal{V}_S}) \geq 1$. Otherwise, $\phi(\mathcal{M}_\mathcal{V}) - \phi(\mathcal{M}_{\mathcal{V}_S}) \geq (f(S) - 1)/2$.

Theorem 7 Let $\mathcal{V}$ be an uncrossable family of bisets. There exist $w: V \to W$ and a spider S activated by w such that

$$\frac{w(V)}{\phi(\mathcal{M}_{\mathcal{V}}) - \phi(\mathcal{M}_{\mathcal{V}_S})} = O(\max\{\gamma, 1\}) \cdot \frac{\text{OPT}}{\phi(\mathcal{M}_{\mathcal{V}})}.$$

Proof Theorem 6 shows that there exist $w \colon V \to W$ and a spider S activated by w such that

$$\frac{w(V)}{f(S)} \leq \frac{2\text{OPT}}{|\mathcal{M}_{\mathcal{V}}|}.$$

Since $\phi(\mathcal{M}_{\mathcal{V}}) \leq (2\gamma + 1)|\mathcal{M}_{\mathcal{V}}|$, we have

$$\frac{w(V)}{f(S)} \leq \frac{2\text{OPT}}{|\mathcal{M}_{\mathcal{V}}|} \leq 2(2\gamma + 1) \cdot \frac{\text{OPT}}{\phi(\mathcal{M}_{\mathcal{V}})}. \tag{10}$$

If $f(S) = 1$, then $\phi(\mathcal{M}_{\mathcal{V}}) - \phi(\mathcal{M}_{\mathcal{V}_S}) \geq f(S)$ by Lemma 7, and hence the required inequality follows from (10). Otherwise, $\phi(\mathcal{M}_{\mathcal{V}}) - \phi(\mathcal{M}_{\mathcal{V}_S}) \geq (f(S) - 1)/2$ by Lemma 7, and hence

$$\frac{w(V)}{f(S)} \geq \frac{w(V)}{2(f(S) - 1)} \geq \frac{w(V)}{4(\phi(\mathcal{M}_{\mathcal{V}}) - \phi(\mathcal{M}_{\mathcal{V}_S}))},$$

where the first inequality follows from $f(S) \geq 2$. Combined with (10), this gives

$$\frac{w(V)}{\phi(\mathcal{M}_{\mathcal{V}}) - \phi(\mathcal{M}_{\mathcal{V}_S})} \leq 8(2\gamma + 1) \cdot \frac{\text{OPT}}{\phi(\mathcal{M}_{\mathcal{V}})}.$$

$\square$

Theorem 8 *Let $\mathcal{V}$ be an uncrossable biset family on a node set V. Let $\gamma = \max_{\hat{X} \in \mathcal{V}} |\Gamma(\hat{X})|$ and $\gamma' = \max\{\gamma, 1\}$. The biset covering problem with $\mathcal{V}$ admits an $O(\gamma' \log(\gamma'|V|))$-approximation algorithm.*

Proof Applying Theorem 7 to $\mathcal{V}'$, we obtain $w \colon V \to W$ and a spider S activated by w such that $w(V)/(\phi(\mathcal{M}_{\mathcal{V}}) - \phi(\mathcal{M}_{\mathcal{V}_S})) = O(\gamma') \cdot \text{OPT}/\phi(\mathcal{M}_{\mathcal{V}'})$. If $\phi(\mathcal{M}_{\mathcal{V}_S}) > 0$, then we apply Theorem 7 to $\mathcal{V}_S$. Let w' and S' be the obtained node weights and spider, respectively. We add edges in S' to S, and increase the weight $w(v)$ by $w'(v)$ for each $v \in V$. We repeat this until $\phi(\mathcal{M}_{\mathcal{V}'_S})$ becomes 0. When the above procedure is completed, we have $w(V) = O(\gamma' \log(\phi(\mathcal{M}_{\mathcal{V}}))) \cdot \text{OPT}$. Since $\phi(\mathcal{M}_{\mathcal{V}}) = O(\gamma'|V|)$, this implies that $w(V) = O(\gamma' \log(\gamma'|V|)) \cdot \text{OPT}$. $\square$

As noted above, $\gamma' \leq k$ when $\mathcal{V}$ arises from the augmentation problem. The biset family arising from the augmentation problem is not necessarily uncrossable. Combined with the decomposition into $O(k^3 \log |V|)$ uncrossable families [15], Theorem 8 presents an $O(k^4 \log |V| \log(k|V|))$-approximation algorithm for the augmentation problem, and an $O(r_{\max}^5 \log |V| \log(r_{\max}|V|))$-approximation algorithm for the network activation problem.

7 Other Applications

7.1 Prize-Collecting Network Design Problems

In the prize-collecting version of network design problems, each pair of nodes that demands a certain connectivity is associated with a non-negative number called a *penalty*. Then, we are allowed to violate the connectivity demand, but if we do so, we have to pay the associated penalty.

In Sect. 4, we have already mentioned that Klein and Ravi's algorithm for the node-weighted Steiner tree problem can be extended to the prize-collecting node-weighted Steiner tree problem because of the LP-based analysis of Guha et al. [5] on the algorithm. Similarly, the algorithms for the SNDP and the network activation problem given in Sects. 5 and 6, respectively, can be extended to the prize-collecting versions of the problems.

For the prize-collecting node-weighted Steiner tree problem, Moss and Rabani [17] also proposed an $O(\log n)$-approximation algorithm. Although this approximation ratio is not better than the one obtained by Klein and Ravi's algorithm, they claimed that their algorithm has a useful property which is known as *Lagrangian multiplier preserving*. However, Könemann, Sadeghian, and Sanità [18] pointed out a flaw in the proof of Moss and Rabani. Instead, Könemann et al. gave another $O(\log n)$-approximation algorithm that possesses the same property. Their algorithm is a sophisticated primal-dual algorithm, and does not seem to use the spider covering framework directly.

The algorithms for the prize-collecting node-weighted Steiner tree problem are used as an important subroutine in the algorithms for some related problems in [5, 17]. This aspect was also investigated by Bateni, Hajiaghayi, and Liaghat [19]. They also proposed a primal-dual algorithm for the prize-collecting node-weighted Steiner forest problem.

7.2 Buy-at-bulk Network Design Problems

In the buy-at-bulk network design problem, a demand pair $\{s_i, t_i\} \in \binom{V}{2}$ in an undirected graph $G = (V, E)$ requires routing r_i units of flow along a path. We have to install cables (or facilities) on edges and nodes in G so that they support the flow requested by the given demanded pairs $\{s_1, t_1\}, \ldots, \{s_d, t_d\}$. Each of the edges and nodes is associated with a monotone concave capacity function $f : \mathbb{R}_+ \to \mathbb{R}_+$, which means that installing cables supporting δ units of flow costs $f(\delta)$. The task of the problem is to install cables so that the total cost is minimized.

Chekuri et al. [20] considered the setting in which both edges and nodes are associated with capacity functions which are possibly distinct for them (i.e., functions f_e and f_v are defined for every edge e and node v). They gave an $O(\log d)$-approximation algorithm for the single-source case, that is, $s_1 = \cdots = s_d$. They used the spider covering framework in this algorithm. They also obtained an $O(\log^4 d)$-approximation algorithm for the multicommodity case, where there is no restriction on the demand pairs. A key concept in this second algorithm is a *junction tree*. The algorithm repeatedly chooses minimum density junction trees rather than spiders. They showed that the problem of finding a minimum density junction tree can be reduced to the single-source case.

7.3 Directed Steiner Tree and Forest Problems

The directed Steiner tree and forest problems are variants of the Steiner tree and forest problems defined over digraphs. In these problems, we are given a digraph $G = (V, E)$ with edge weights. In the directed Steiner tree problem, we are also given a root $r \in V$ and a set of terminals $T \subseteq V \setminus \{r\}$, and are required to find a minimum weight subgraph of G that includes a directed path from r to each terminal in T. In the directed Steiner forest problem, we are given a set of node pairs $(s_1, t_1), \ldots, (s_d, t_d)$, and a solution have to include a directed path from s_i to t_i for each $i \in [d]$. We can also consider node weights in these problems, but problem with node weights and those with edge weights are equivalent in digraphs because they can be reduced each other.

While the undirected Steiner tree and forest problems admit constant-factor approximation algorithms, the directed Steiner tree and forest problems are much harder than the corresponding undirected variants. Indeed, the directed Steiner tree problem includes the node-weighted Steiner tree and the group Steiner tree problems in undirected graphs, which seem to admit no sub-polylogarithmic approximation [21].

For the directed Steiner tree problem, Charikar et al. [22] gave an $O(|T|^\epsilon)$-approximation polynomial-time algorithm for any fixed $\epsilon > 0$. They also gave an $O(d^{2/3} \log^{1/3} d)$-approximation polynomial-time algorithm for the directed Steiner forest problem. Their algorithms use an idea of selecting low density subgraphs repeatedly. Subsequently, the approximation ratio for the directed Steiner forest problem was improved by Chekuri et al. [23] to $O(d^{1/2+\epsilon})$ and Feldman, Kortsarz, and Nutov [24] to $O(n^\epsilon \min\{n^{4/5}, |E|^{2/3}\})$. These are based on the idea of junction trees, which was introduced by Chekuri et al. [20] for the buy-at-bulk network design problem.

7.4 Online Network Design Problems

In the online Steiner tree problem, we have a graph $G = (V, E)$ in advance, and nodes are specified as terminals one by one. When a node v is specified as a terminal, an online algorithm is required to connect it immediately to the nodes which have previously been specified as terminals by choosing edges. The performance of the algorithm is measured by the competitive ratio, which is the worst case ratio of the weights of the solution constructed by the algorithm to the minimum cost of the Steiner tree. The problem can be defined both for the edge and node weights.

For the edge-weighted case with a terminal set T, Imase and Waxman [25] proved that a competitive ratio of $O(\log |T|)$ is achieved by a greedy algorithm, which simply connects a terminal to one of the previous terminals via a shortest path. Naor, Panigrahi, and Singh [26] extended this to the node-weighted problem. Although the simple greedy algorithm does not work for the node-weighted Steiner tree, they showed that the problem can be reduced to the online facility location problem, which results in an $O(\log n \log^2 |T|)$-competitive algorithm. In this analysis, the spider decomposition theorem due to Klein and Ravi [2] played an important role.

In addition to the abovementioned research, Hajiaghayi, Liaghat, and Panigrahi [27] gave an online algorithm for the node-weighted Steiner forest. Finally, Ene et al. [28] gave an online algorithm for the buy-at-bulk network design problem.

8 Conclusion

In this article, we surveyed algorithms given by the spider covering framework. This framework extends the well-known greedy algorithm for the set cover problem to network settings. It is useful for many network design problems such as node-weighted problems and network activation problems, which are known as relatively difficult problems because they include the set cover problem. We believe that this framework is a strong tool for tackling difficult network design problems.

References

1. P.N. Klein, R. Ravi, A nearly best-possible approximation algorithm for node-weighted Steiner trees, in *Proceedings of the 3rd Integer Programming and Combinatorial Optimization Conference*, Erice, Italy, 29 April–1 May (1993), pp. 323–332
2. P.N. Klein, R. Ravi, A nearly best-possible approximation algorithm for node-weighted Steiner trees. J. Algorithms **19**(1), 104–115 (1995)
3. J. Byrka, F. Grandoni, T. Rothvoß, L. Sanità, Steiner tree approximation via iterative randomized rounding. J. ACM **60**(1), 6 (2013)
4. M.X. Goemans, N. Olver, T. Rothvoß, R. Zenklusen, Matroids and integrality gaps for hypergraphic steiner tree relaxations, in *Proceedings of the 44th Symposium on Theory of Computing Conference, STOC*, New York, NY, USA, 19 - 22 May (2012), pp. 1161–1176
5. S. Guha, A. Moss, J. Naor, B. Schieber, Efficient recovery from power outage (extended abstract), in *Proceedings of the Thirty-First Annual ACM Symposium on Theory of Computing*, Atlanta, Georgia, USA, 1-4 May (1999), pp. 574–582
6. Z. Nutov, Approximating Steiner networks with node-weights. SIAM J. Comput. **39**(7), 3001–3022 (2010)
7. Z. Nutov, Approximating minimum-cost connectivity problems via uncrossable bifamilies. ACM Trans. Algorithms **9**(1), 1 (2012)
8. Z. Nutov, Survivable network activation problems. Theor. Comput. Sci. **514**, 105–115 (2013)

9. T. Fukunaga, Covering problems in edge- and node-weighted graphs. Disc. Opt. **20**, 40–61 (2016)
10. D. Bienstock, M.X. Goemans, D. Simchi-Levi, D.P. Williamson, A note on the prize collecting traveling salesman problem. Math. Program. **59**, 413–420 (1993)
11. C. Chekuri, A. Ene, A. Vakilian, Prize-collecting survivable network design in node-weighted graphs, *APPROX-RANDOM. Lecture Notes in Computer Science*, vol. 7408 (2012), pp. 98–109
12. T. Fukunaga, Spider covers for prize-collecting network activation problem, in *Proceedings of the Twenty-Sixth Annual ACM-SIAM Symposium on Discrete Algorithms, SODA 2015*, San Diego, CA, USA, 4-6 Jan (2015), pp. 9–24
13. D. Panigrahi, Survivable network design problems in wireless networks, in *Proceedings of the Twenty-Second Annual ACM-SIAM Symposium on Discrete Algorithms, SODA 2011*, San Francisco, CA, USA, 23–25 Jan (2011), pp. 1014–1027
14. A. Vakilian, Node-weighted prize-collecting survivable network design problems. Master's thesis, University of Illinois at Urbana-Champaign, 2013
15. J. Chuzhoy, S. Khanna, An $O(k^3 \log n)$-approximation algorithm for vertex-connectivity survivable network design. Theory Comput. **8**(1), 401–413 (2012)
16. Z. Nutov, Approximating subset k-connectivity problems. J. Discret. Algorithms **17**, 51–59 (2012)
17. A. Moss, Y. Rabani, Approximation algorithms for constrained node weighted Steiner tree problems. SIAM J. Comput. **37**(2), 460–481 (2007)
18. J. Könemann, S.S. Sadeghabad, L. Sanità, An LMP $O(\log n)$-approximation algorithm for node weighted prize collecting Steiner tree, in *54th Annual IEEE Symposium on Foundations of Computer Science, FOCS*, Berkeley, CA, USA, 26–29 Oct (2013), pp. 568–577
19. M.H. Bateni, M.T. Hajiaghayi, V. Liaghat, Improved approximation algorithms for (budgeted) node-weighted Steiner problems, in *Automata, Languages, and Programming - 40th International Colloquium, ICALP 2013*, Riga, Latvia, 8–12 July 2013, Proceedings, Part I. Lecture Notes in Computer Science, vol. 7965 (2013), pp. 81–92
20. C. Chekuri, M.T. Hajiaghayi, G. Kortsarz, M.R. Salavatipour, Approximation algorithms for nonuniform buy-at-bulk network design. SIAM J. Comput. **39**(5), 1772–1798 (2010)
21. E. Halperin, R. Krauthgamer, Polylogarithmic inapproximability, in *Proceedings of the 35th Annual ACM Symposium on Theory of Computing*, San Diego, CA, USA, June 9-11 (2003), pp. 585–594
22. M. Charikar, C. Chekuri, T.-Y. Cheung, Z. Dai, A. Goel, S. Guha, M. Li, Approximation algorithms for directed Steiner problems. J. Algorithms **33**(1), 73–91 (1999)
23. C. Chekuri, G. Even, A. Gupta, D. Segev, Set connectivity problems in undirected graphs and the directed steiner network problem. ACM Trans. Algorithms **7**(2), 18 (2011)
24. M. Feldman, G. Kortsarz, Z. Nutov, Improved approximation algorithms for directed Steiner forest. J. Comput. Syst. Sci. **78**(1), 279–292 (2012)
25. M. Imase, B.M. Waxman, Dynamic steiner tree problem. SIAM J. Discrete Math. **4**(3), 369–384 (1991)
26. J. Naor, D. Panigrahi, M. Singh, Online node-weighted Steiner tree and related problems, in *IEEE 52nd Annual Symposium on Foundations of Computer Science, FOCS 2011*, Palm Springs, CA, USA, 22–25 Oct (2011) pp. 210–219
27. M.T. Hajiaghayi, V. Liaghat, D. Panigrahi, Online node-weighted steiner forest and extensions via disk paintings. . SIAM J. Comput. **46**(3), 911–935 (2017)
28. A. Ene, D. Chakrabarty, R. Krishnaswamy, D. Panigrahi, Online buy-at-bulk network design, in *IEEE 56th Annual Symposium on Foundations of Computer Science, FOCS 2015, Berkeley*, CA, USA, 17-20 Oct (2015), pp. 545–562

Discrete Convex Functions on Graphs and Their Algorithmic Applications

Hiroshi Hirai

Abstract The present article is an exposition of a theory of discrete convex functions on certain graph structures, developed by the author in recent years. This theory is a spin-off of discrete convex analysis by Murota, and is motivated by combinatorial dualities in multiflow problems and the complexity classification of facility location problems on graphs. We outline the theory and algorithmic applications in combinatorial optimization problems.

1 Introduction

The present article is an exposition of a theory of discrete convex functions on certain graph structures, developed by the author in recent years. This theory is viewed as a spin-off of *Discrete Convex Analysis (DCA)*, which is a theory of convex functions on the integer lattice and has been developed by Murota and his collaborators in the last 20 years; see [43–45] and [12, Chap. VII]. Whereas the main targets of DCA are matroid-related optimization (or submodular optimization), our new theory is motivated by combinatorial dualities arising from multiflow problems [23] and the complexity classification of certain facility location problems on graphs [27].

The heart of our theory is analogues of $L^\natural$-*convex functions* [10, 16] for certain graph structures, where $L^\natural$-convex functions are one of the fundamental classes of discrete convex functions on $\mathbf{Z}^n$, and play primary roles in DCA. These analogues are inspired by the following intriguing properties of $L^\natural$-convex functions:

- An $L^\natural$-convex function is (equivalently) defined as a function g on $\mathbf{Z}^n$ satisfying a discrete version of the convexity inequality, called the *discrete midpoint convexity*:

$$g(x) + g(y) \geq g(\lfloor (x+y)/2 \rfloor) + g(\lceil (x+y)/2 \rceil) \quad (x, y \in \mathbf{Z}^n), \qquad (1)$$

H. Hirai (✉)
Department of Mathematical Informatics, Graduate School of Information Science
and Technology, The University of Tokyo, Tokyo 113-8656, Japan
e-mail: hirai@mist.i.u-tokyo.ac.jp

© Springer Nature Singapore Pte Ltd. 2017
T. Fukunaga and K. Kawarabayashi (eds.), *Combinatorial Optimization
and Graph Algorithms*, DOI 10.1007/978-981-10-6147-9_4

where $\lfloor \cdot \rfloor$ and $\lceil \cdot \rceil$ are operators rounding down and up, respectively, the fractional part of each component.

- $L^{\natural}$-convex function g is *locally submodular* in the following sense: For each $x \in \mathbf{Z}^n$, the function on $\{0, 1\}^n$ defined by $u \mapsto g(x + u)$ is submodular.
- Analogous to ordinary convex functions, $L^{\natural}$-convex function g enjoys an optimality criterion of a local-to-global type: If x is not a minimizer of g, then there exists $y \in (x + \{0, 1\}^n) \cup (x - \{0, 1\}^n)$ with $g(y) < g(x)$.
- This leads us to a conceptually-simple minimization algorithm, called the *steepest descent algorithm (SDA)*: For $x \in \mathbf{Z}^n$, find a (local) minimizer y of g over $(x + \{0, 1\}^n) \cup (x - \{0, 1\}^n)$ via submodular function minimization (SFM). If $g(y) = g(x)$, then x is a (global) minimizer of g. If $g(y) < g(x)$, then let $x := y$, and repeat.
- The number of iterations of SDA is sharply bounded by a certain l_∞-distance between the initial point and minimizers [46].
- $L^{\natural}$-convex function g is extended to a convex function $\overline{g}$ on $\mathbf{R}^n$ via the *Lovász extension*, and this convexity property characterizes the $L^{\natural}$-convexity.

Details are given in [44], and a recent survey [49] is also a good source of $L^{\natural}$-convex functions.

We consider certain classes of graphs Γ that canonically define functions (on the vertex set of Γ) having analogous properties, which we call *L-convex functions on* Γ (with $^{\natural}$ omitted). The aim of this paper is to explain these L-convex functions and their roles in combinatorial optimization problems, and to demonstrate the SDA-based algorithm design.

Our theory is parallel with recent developments in generalized submodularity and *valued constraint satisfaction problem (VCSP)* [39, 51, 54]. Indeed, localizations of these L-convex functions give rise to a rich subclass of generalized submodular functions that include k-submodular functions [29] and submodular functions on diamonds [15, 41], and are polynomially minimizable in the VCSP setting.

The starting point is the observation that if $\mathbf{Z}^n$ is viewed as a grid graph (with order information), some of the above properties of $L^{\natural}$-convex functions are still well-formulated. Indeed, extending the discrete midpoint convexity to trees, Kolmogorov [38] introduced a class of discrete convex functions, called *tree-submodular functions*, on the product of rooted trees, and showed several analogous results. In Sect. 2, we discuss L-convex functions on such grid-like structures. We start by considering a variation of tree-submodular functions, where the underlying graph is the product of zigzagly-oriented trees. Then we explain that a theory of L-convex functions is naturally developed on a structure, known as *Euclidean building* [1], which is a kind of an amalgamation of $\mathbf{Z}^n$ and is a far-reaching generalization of a tree. Applications of these L-convex functions are given in Sect. 3. We outline SDA-based efficient combinatorial algorithms for two important multiflow problems. In Sect. 4, we explain L-convex functions on a more general class of graphs, called *oriented modular graphs*. This graph class emerged from the complexity classification of the *minimum 0-extension problem* [34]. We outline that the theory of L-convex functions leads to a solution of this classification problem. This was the original motivation of our theory.

The contents of this paper are based on the results in papers [25–28], in which further details and omitted proofs are found.

Let $\mathbf{R}$, $\mathbf{R}_+$, $\mathbf{Z}$, and $\mathbf{Z}_+$ denote the sets of reals, nonnegative reals, integers, and nonnegative integers, respectively. In this paper, $\mathbf{Z}$ is often regarded as an infinite path obtained by adding an edge to each consecutive pair of integers. Let $\overline{\mathbf{R}} := \mathbf{R} \cup \{\infty\}$, where ∞ is the infinity element treated as $a < \infty$, $a + \infty = \infty$ for $a \in \mathbf{R}$, and $\infty + \infty = \infty$. For a function $g : X \to \overline{\mathbf{R}}$ on a set X, let dom $g := \{x \in X \mid g(x) < \infty\}$.

2 L-Convex Function on Grid-Like Structure

In this section, we discuss L-convex functions on grid-like structures. In the first two subsections (Sects. 2.1 and 2.2), we consider specific underlying graphs (tree-product and twisted tree-product) that admit analogues of discrete midpoint operators and the corresponding L-convex functions. In both cases, the local submodularity and the local-to-global optimality criterion are formulated in a straightforward way (Lemmas 2.1, 2.2, 2.3, and 2.4). In Sect. 2.3, we introduce the steepest descent algorithm (SDA) in a generic way, and present the iteration bound (Theorem 2.5). In Sect. 2.4, we explain that the L-convexity is naturally generalized to that in Euclidean buildings of type C, and that the Lovász extension theorem can be generalized via the geometric realization of Euclidean buildings and CAT(0)-metrics (Theorem 2.6).

We use a standard terminology on posets (partially ordered sets) and lattices; see e.g., [21]. Let $P = (P, \preceq)$ be a poset. For an element $p \in P$, the *principal ideal* I_p is the set of elements $q \in P$ with $q \preceq p$, and the *principal filter* F_p is the set of elements $q \in P$ with $p \preceq q$. For $p, q \in P$ with $p \preceq q$, let $\max\{p, q\} := q$ and $\min\{p, q\} := p$. A *chain* is a subset $X \subseteq P$ such that for every $p, q \in X$ it holds that $p \preceq q$ or $q \preceq p$. For $p \preceq q$, the *interval* $[p, q]$ of p, q is defined as $[p, q] := \{u \in P \mid p \preceq u \preceq q\}$. For two posets P, P', the direct product $P \times P'$ becomes a poset by the direct product order $\preceq$ defined by $(p, p') \preceq (q, q')$ if and only if $p \preceq q$ in P and $p' \preceq q'$ in P'.

For an undirected graph G, an edge joining vertices x and y is denoted by xy. When G plays a role of the domain of discrete convex functions, the vertex set $V(G)$ of G is also denoted by G. Let $d = d_G$ denote the shortest path metric on G. We often endow G with an edge-orientation, where $x \to y$ means that edge xy is oriented from x to y. For two graphs G and H, let $G \times H$ denote the Cartesian product of G and H, i.e., the vertices are all pairs of vertices of G and H and two vertices (x, y) and (x', y') are adjacent if and only if $d_G(x, x') + d_H(y, y') = 1$.

2.1 L-Convex Function on Tree-Grid

Let G be a tree. Let B and W denote the color classes of G viewed as a bipartite graph. Endow G with a zigzag orientation so that $u \to v$ if and only if $u \in W$ and $v \in B$. This orientation is acyclic. The induced partial order on G is denoted by $\preceq$, where $v \leftarrow u$ is interpreted as $v \preceq u$.

Discrete midpoint operators $\bullet$ and $\circ$ on G are defined as follows. For vertices $u, v \in G$, there uniquely exists a pair (a, b) of vertices such that $d(u, v) = d(u, a) + d(a, b) + d(b, v), d(u, a) = d(b, v)$, and $d(a, b) \leq 1$. In particular, a and b are equal or adjacent, and hence comparable. Let $u \bullet v := \min\{a, b\}$ and $u \circ v := \max\{a, b\}$.

Consider the product G^n of G; see Fig. 1 for G^2. The operators $\bullet$ and $\circ$ are extended on G^n component-wise: For $x = (x_1, x_2, \ldots, x_n), y = (y_1, y_2, \ldots, y_n) \in G^n$, let $x \bullet y$ and $x \circ y$ be defined by

$$x \bullet y := (x_1 \bullet y_1, x_2 \bullet y_2, \ldots, x_n \bullet y_n), \quad x \circ y := (x_1 \circ y_1, x_2 \circ y_2, \ldots, x_n \circ y_n).$$
(2)

Mimicking the discrete midpoint convexity (1), a function $g : G^n \to \overline{\mathbf{R}}$ is called *L-convex* (or *alternating L-convex* [25]) if it satisfies

$$g(x) + g(y) \geq g(x \bullet y) + g(x \circ y) \quad (x, y \in G^n).$$
(3)

By this definition, a local-to-global optimality criterion is obtained in a straightforward way. Recall the notation that I_x and F_x are the principal ideal and filter of vertex $x \in G^n$.

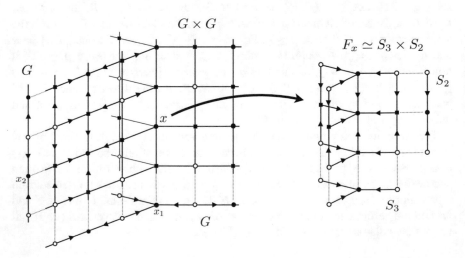

Fig. 1 Tree-grid G^2. *Black* and *white* points in G represent vertices in B and W, respectively, The principal filter of a vertex $x \in G^2$ is picked out to the *right*, which is a semilattice isomorphic to $S_3 \times S_2$

Lemma 2.1 ([25]) *Let* $g : G^n \to \overline{\mathbf{R}}$ *be an L-convex function. If* $x \in \mathrm{dom}\, g$ *is not a minimizer of* g, *there exists* $y \in I_x \cup F_x$ *with* $g(y) < g(x)$.

Proof Suppose that $x \in \mathrm{dom}\, g$ is not a minimizer of g. There is $z \in \mathrm{dom}\, g$ such that $g(z) < g(x)$. Choose such z with minimum $\max_i d(x_i, z_i)$. By $g(x) + g(z) \geq g(x \bullet z) + g(x \circ z)$, it necessarily holds that $\max_i d(x_i, z_i) \leq 1$, and one of $x \bullet z \in I_x$ and $x \circ z \in F_x$ has a smaller value of g than x. ∎

This lemma says that I_x and F_x are "neighborhoods" of x for which the local optimality is defined. This motivates us to consider the class of functions appearing as the restrictions of g to I_x and F_x for $x \in G^n$, which we call the *localizations* of g. The localizations of L-convex functions give rise to a class of submodular-type discrete convex functions known as *k-submodular functions* [29]. To explain this fact, we introduce a class of semilattices isomorphic to I_x or F_x.

For a nonnegative integer k, let S_k denote a $(k+1)$-element set with a special element 0. Define a partial order $\preceq$ on S_k by $0 \preceq u$ for $u \in S_k \setminus \{0\}$ with no other relations. Let $\sqcup$ and $\sqcap$ be binary operations on S_k defined by

$$u \sqcap v := \begin{cases} \min\{u, v\} & \text{if } u \preceq v \text{ or } v \preceq u, \\ 0 & \text{otherwise,} \end{cases} \quad u \sqcup v := \begin{cases} \max\{u, v\} & \text{if } u \preceq v \text{ or } v \preceq u, \\ 0 & \text{otherwise.} \end{cases}$$

For an n-tuple $\mathbf{k} = (k_1, k_2, \ldots, k_n)$ of nonnegative integers, let $S_{\mathbf{k}} := S_{k_1} \times S_{k_2} \times \cdots \times S_{k_n}$. A function $f : S_{\mathbf{k}} \to \overline{\mathbf{R}}$ is *$\mathbf{k}$-submodular* if it satisfies

$$f(x) + f(y) \geq f(x \sqcap y) + f(x \sqcup y) \quad (x, y \in S_{\mathbf{k}}), \tag{4}$$

where operators $\sqcap$ and $\sqcup$ are extended to $S_{\mathbf{k}}$ component-wise. Then $\mathbf{k}$-submodular functions are identical to submodular functions if $\mathbf{k} = (1, 1, \ldots, 1)$ and to bisubmodular functions if $\mathbf{k} = (2, 2, \ldots, 2)$.

Let us return to tree-product G^n. For every point $x \in G^n$, the principal filter F_x is isomorphic to $S_{\mathbf{k}}$ for $\mathbf{k} = (k_1, k_2, \ldots, k_n)$, where $k_i = 0$ if $x_i \in W$ and k_i is equal to the degree of x_i in G if $x_i \in B$. Similarly for the principal ideal I_x (with partial order reversed).

Observe that operations $\bullet$ and $\circ$ coincides with $\sqcap$ and $\sqcup$ (resp. $\sqcup$ and $\sqcap$) in any principle filter (resp. ideal). Then an L-convex function on G^n is locally $\mathbf{k}$-submodular in the following sense.

Lemma 2.2 ([25]) *An L-convex function* $g : G^n \to \overline{\mathbf{R}}$ *is* $\mathbf{k}$-*submodular on* F_x *and on* I_x *for every* $x \in \mathrm{dom}\, g$.

In particular, an L-convex function on a tree-grid can be minimized via successive $\mathbf{k}$-submodular function minimizations; see Sect. 2.3.

2.2 L-Convex Function on Twisted Tree-Grid

Next we consider a variant of a tree-grid, which is obtained by *twisting* the product of two trees, and by taking the product. Let G and H be infinite trees without vertices of degree one. Consider the product $G \times H$, which is also a bipartite graph. Let B and W denote the color classes of $G \times H$. For each 4-cycle C of $G \times H$, add a new vertex w_C and new four edges joining w_C and vertices in C. Delete all original edges of $G \times H$, i.e., all edges non-incident to new vertices. The resulting graph is denoted by $G \boxtimes H$. Endow $G \boxtimes H$ with an edge-orientation such that $x \to y$ if and only if $x \in W$ or $y \in B$. Let $\preceq$ denote the induced partial order on $G \boxtimes H$. See Fig. 2 for $G \boxtimes H$.

Discrete midpoint operators $\bullet$ and $\circ$ are defined as follows. Consider two vertices x, y in $G \boxtimes H$. Then $x \in C$ or $x = w_C$ for some 4-cycle C in $G \times H$. Similarly, $y \in D$ or $y = w_D$ for some 4-cycle D in $G \times H$. There are infinite paths P in G and Q in H such that $P \times Q$ contains both C and D. Indeed, consider the projections of C and D to G, which are edges. Since G is a tree, we can choose P as an infinite path in G containing both of them. Similarly for Q. Now the subgraph $P \boxtimes Q$ of $G \boxtimes H$ coincides with (45-degree rotation of) the product of two paths with the zigzag orientation as in Sect. 2.1. Thus, in $P \boxtimes Q$, the discrete midpoint points $x \bullet y$ and $x \circ y$ are defined. They are determined independently of the choice of P and Q.

Consider the n-product $(G \boxtimes H)^n$ of $G \boxtimes H$, which is called a *twisted tree-grid*. Similarly to the previous case, an *L-convex function* on $(G \boxtimes H)^n$ is defined as a function $g : (G \boxtimes H)^n \to \overline{\mathbf{R}}$ satisfying

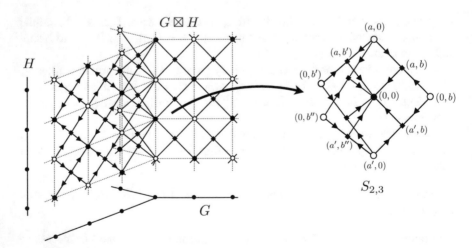

Fig. 2 Twisted tree-grid $G \boxtimes H$. *Dotted lines* represent edges of $G \times H$. *Black* and *white* (round) points represent vertices in B and W, respectively, while square points correspond to 4-cycles in $G \times H$. The principal filter of a *black* vertex is picked out to the *right*, which is a semilattice isomorphic to $S_{2,3}$

$$g(x) + g(y) \geq g(x \bullet y) + g(x \circ y) \quad (x, y \in (G \boxtimes H)^n), \tag{5}$$

where operations $\bullet$ and $\circ$ are extended on $(G \boxtimes H)^n$ component-wise as before. Again the following holds, where the proof is exactly the same as Lemma 2.1.

Lemma 2.3 *Let $g : (G \boxtimes H)^n \to \overline{\mathbf{R}}$ be an L-convex function. If $x \in \mathrm{dom}\, g$ is not a minimizer of g, there is $y \in I_x \cup F_x$ with $g(y) < g(x)$.*

As before, we study localizations of an L-convex function on $(G \boxtimes H)^n$. This naturally leads to further generalizations of k-submodular functions. For positive integers k, l, consider product $S_k \times S_l$ of S_k and S_l. Define a partial order $\preceq'$ on $S_k \times S_l$ by $(0, 0) \preceq' (a, b) \preceq' (a, 0)$ and $(a, b) \preceq' (0, b)$ for $(a, b) \in S_k \times S_l$ with $a \neq 0$ and $b \neq 0$. The resulting poset is denoted by $S_{k,l}$; note that $\preceq'$ is different from the direct product order.

The operations $\sqcup$ and $\sqcap$ on $S_{k,l}$ are defined as follows. For $p, q \in S_{k,l}$, there are $a, a' \in S_k \setminus \{0\}$ and $b, b' \in S_l \setminus \{0\}$ such that $p, q \in \{0, a, a'\} \times \{0, b, b'\}$. The restriction of $S_{k,l}$ to $\{0, a, a'\} \times \{0, b, b'\}$ is isomorphic to $S_2 \times S_2 = \{-1, 0, 1\}^2$ (with order $\preceq$), where an isomorphism φ is given by

$(0, 0) \mapsto (0, 0),$

$(a, b) \mapsto (1, 0), (a', b) \mapsto (0, -1), (a, b') \mapsto (0, 1), (a', b') \mapsto (-1, 0),$

$(a, 0) \mapsto (1, 1), (a', 0) \mapsto (-1, -1), (0, b) \mapsto (1, -1), (0, b') \mapsto (-1, 1).$

Then $p \sqcup q := \varphi^{-1}(\varphi(p) \sqcup \varphi(q))$ and $p \sqcap q := \varphi^{-1}(\varphi(p) \sqcap \varphi(q))$. Notice that they are determined independently of the choice of a, a', b, b' and φ.

For a pair of n-tuples $\boldsymbol{k} = (k_1, k_2, \ldots, k_n)$ and $\boldsymbol{l} = (l_1, l_2, \ldots, l_n)$, let $S_{\boldsymbol{k}, \boldsymbol{l}} := S_{k_1, l_1} \times S_{k_2, l_2} \times \cdots \times S_{k_n, l_n}$. A function $f : S_{\boldsymbol{k}, \boldsymbol{l}} \to \overline{\mathbf{R}}$ is $(\boldsymbol{k}, \boldsymbol{l})$-*submodular* if it satisfies

$$f(x) + f(y) \geq f(x \sqcap y) + f(x \sqcup y) \quad (x, y \in S_{\boldsymbol{k}, \boldsymbol{l}}). \tag{6}$$

Poset $S_{\boldsymbol{k}, \boldsymbol{l}}$ contains $S_{0, l} \simeq S_l$ and *diamonds* (i.e., a modular lattice of rank 2) as $(\sqcap, \sqcup)$-closed sets. Thus $(\boldsymbol{k}, \boldsymbol{l})$-submodular functions can be viewed as a common generalization of $\boldsymbol{k}$-submodular functions and submodular functions on diamonds.

Both the principal ideal and filter of each vertex in $(G \boxtimes H)^n$ are isomorphic to $S_{\boldsymbol{k}, \boldsymbol{l}}$ for some $\boldsymbol{k}, \boldsymbol{l}$, in which $\{\bullet, \circ\}$ are equal to $\{\sqcap, \sqcup\}$. Thus we have:

Lemma 2.4 *An L-convex function $g : (G \boxtimes H)^n \to \overline{\mathbf{R}}$ is $(\boldsymbol{k}, \boldsymbol{l})$-submodular on I_x and on F_x for every $x \in \mathrm{dom}\, g$.*

2.3 Steepest Descent Algorithm

The two classes of L-convex functions in the previous subsection can be minimized by the same principle, analogous to the steepest descent algorithm for $L^\natural$-convex functions. Let Γ be a tree-product G^n or a twisted tree-product $(G \boxtimes H)^n$. We here consider L-convex functions g on Γ such that a minimizer of g exists.

Steepest Descent Algorithm (SDA)

Input: An L-convex function $g : \Gamma \to \overline{\mathbf{R}}$ and an initial point $x \in \text{dom } g$.
Output: A minimizer x of g.
Step 1:   Find a minimizer y of g over $I_x \cup F_x$.
Step 2:   If $g(x) = g(y)$, then output x and stop; x is a minimizer.
Step 3:   Let $x := y$, and go to step 1.

The correctness of this algorithm follows immediately from Lemmas 2.1 and 2.3. A minimizer y in Step 1 is particularly called a *steepest direction* at x. Notice that a steepest direction is obtained by minimizing g over I_x and over F_x, which are k- or (k, l)-submodular function minimization by Lemmas 2.2 and 2.4.

Besides its conceptual simplicity, there remain two issues in applying SDA to specific combinatorial optimization problems:

- How to minimize g over I_x and over F_x.
- How to estimate the number of iterations.

We first discuss the second issue. In fact, there is a surprisingly simple and sharp iteration bound, analogous to the case of $L^\natural$-convex functions [46]. If Γ is a tree-grid G^n, then define Γ^Δ as the graph obtained from Γ by adding an edge to each pair of (distinct) vertices x, y with $d(x_i, y_i) \leq 1$ for $i = 1, 2, \ldots, n$. If Γ is a twisted tree-grid $(G \boxtimes H)^n$, then define Γ^Δ as the graph obtained from Γ by adding an edge to each pair of (distinct) vertices x, y such that x_i and y_i belong to a common 4-cycle in $G \boxtimes H$ for each $i = 1, 2, \ldots, n$. Let $d_\Delta := d_{\Gamma^\Delta}$. Observe that x and $y \in I_x \cup F_x$ are adjacent in Γ^Δ. Hence the number of iterations of SDA is at least the minimum distance $d_\Delta(x, \text{opt}(g)) := \min\{d_\Delta(x, y) \mid y \in \text{opt}(g)\}$ from the initial point x and the minimizer set $\text{opt}(g)$ of g. This lower bound is almost tight.

Theorem 2.5 ([25, 26]) *The number of the iterations of SDA applied to L-convex function g and initial point $x \in \text{dom } g$ is at most $d_\Delta(x, \text{opt}(g)) + 2$.*

For specific problems (considered in Sect. 3), the upper bound of $d_\Delta(x, \text{opt}(g))$ is relatively easier to be estimated.

Thus we concentrate only on the first issue. Minimizing g over I_x and F_x is a k-submodular function minimization if Γ is a tree-grid and is a (k, l)-submodular function minimization if Γ is a twisted tree-grid. Currently no polynomial time algorithm is known for k-submodular function minimization under the oracle model. However, under several important special cases (including VCSP model), the above submodular functions can be minimizable in polynomial time, and SDA is implementable; see Sect. 4. Moreover, for further special classes of k- or (k, l)-submodular functions arising from our target problems in Sect. 3, a fast and efficient minimization via network or submodular flows is possible, and hence the SDA framework brings efficient combinatorial polynomial time algorithms.

2.4 L-Convex Function on Euclidean Building

The arguments in the previous sections are naturally generalized to the structures known as *spherical* and *Euclidean buildings of type C*; see [1, 52] for the theory of buildings. We here explain a building-theoretic approach to L-convexity and submodularity. We start with a local theory; we introduce a *polar space* and *submodular functions* on it. Polar spaces are equivalent to spherical buildings of type C [52], and generalize domains S_k and $S_{k,l}$ for k-submodular and (k, l)-submodular functions.

A polar space L of rank n is defined as a poset endowed with a system of subposets, called *polar frames*, satisfying the following axioms:

P0: Each polar frame is isomorphic to S_2^n.

P1: For two chains C, D in L, there is a polar frame F containing them.

P2: If polar frames F, F' both contain two chains C, D, then there is an isomorphism $F \to F'$ being identity on C and D.

Namely a polar space is viewed as an amalgamation of several S_2^n. Observe that S_k and $S_{k,l}$ (with $k, l \geq 2$) are polar spaces, and so are their products.

Operators $\sqcap$ and $\sqcup$ on polar space L are defined as follows. For x, $y \in L$, consider a polar frame $F \simeq S_2^n$ containing x, y (via P1), and $x \sqcap y$ and $x \sqcup y$ can be defined in F. One can derive from the axioms that $x \sqcap y$ and $x \sqcup y$ are determined independently of the choice of a polar frame F. Thus operators $\sqcap$ and $\sqcup$ are well-defined.

A *submodular function* on a polar space L is a function $f : L \to \mathbf{R}$ satisfying

$$f(x) + f(y) \geq f(x \sqcap y) + f(x \sqcup y) \quad (x, y \in L). \tag{7}$$

Equivalently, a submodular function on a polar space is a function being bisubmodular on each polar frame.

Next we introduce a *Euclidean building (of type C)* and *L-convex functions* on it. A Euclidean building is simply defined from the above axioms by replacing S_2^n by $\mathbf{Z}^n$, where $\mathbf{Z}$ is zigzagly ordered as

$$\cdots \succ -2 \prec -1 \succ 0 \prec 1 \succ 2 \prec \cdots .$$

Namely a Euclidean building of rank n is a poset Γ endowed with a system of subposets, called *apartments*, satisfying:

B0: Each apartment is isomorphic to $\mathbf{Z}^n$.

B1: For any two chains A, B in Γ, there is an apartment Σ containing them.

B2: If Σ and Σ' are apartments containing two chains A, B, then there is an isomorphism $\Sigma \to \Sigma'$ being identity on A and B.

A tree-product G^n and twisted tree-product $(G \boxtimes H)^n$ are Euclidean buildings (of rank n and $2n$, respectively). In the latter case, apartments are given by $(P_1 \boxtimes Q_1) \times (P_2 \boxtimes Q_2) \times \cdots \times (P_n \boxtimes Q_n) \simeq \mathbf{Z}^{2n}$ for infinite paths P_i in G and Q_i in H for $i = 1, 2, \ldots, n$.

The discrete midpoint operators are defined as follows. For two vertices $x, y \in \Gamma$, choose an apartment Σ containing x, y. Via isomorphism $\Sigma \simeq \mathbf{Z}^n$, discrete midpoints $x \bullet y$ and $x \circ y$ are defined as in the previous section. Again, $x \bullet y$ and $x \circ y$ are independent of the choice of apartments.

An *L-convex function* on Γ is a function $g : \Gamma \to \overline{\mathbf{R}}$ satisfying

$$g(x) + g(y) \geq g(x \bullet y) + g(x \circ y) \quad (x, y \in \Gamma). \tag{8}$$

Observe that each principal ideal and filter of Γ are polar spaces. Then the previous Lemmas 2.1, 2.2, 2.3, 2.4, and Theorem 2.5 are generalized as follows:

- An L-convex function $g : \Gamma \to \overline{\mathbf{R}}$ is submodular on polar spaces I_x and F_x for every $x \in \Gamma$.
- If x is not a minimizer of g, then there is $y \in I_x \cup F_x$ with $g(y) < g(x)$. In particular, the steepest descent algorithm (SDA) is well-defined, and correctly obtains a minimizer of g (if it exists).
- The number of iterations of SDA for g and initial point $x \in \mathrm{dom}\, g$ is bounded by $d_\Delta(x, \mathrm{opt}(g)) + 2$, where $d_\Delta(y, z)$ is the l_∞-distance between y and z in the apartment $\Sigma \simeq \mathbf{Z}^n$ containing y, z.

Next we discuss a convexity aspect of L-convex functions. Analogously to $\mathbf{Z}^n \hookrightarrow \mathbf{R}^n$, there is a continuous metric space $K(\Gamma)$ into which Γ is embedded. Accordingly, any function g on Γ is extended to a function $\bar{g}$ on $K(\Gamma)$, which is an analogue of the Lovász extension. It turns out that the L-convexity of g is equivalent to the convexity of $\bar{g}$ with respect to the metric on $K(\Gamma)$.

We explain this fact more precisely. Let $K(\Gamma)$ be the set of formal sums

$$\sum_{p \in \Gamma} \lambda(p) p$$

of vertices in Γ for $\lambda : \Gamma \to \mathbf{R}_+$ satisfying that $\{p \in \Gamma \mid \lambda(p) \neq 0\}$ is a chain and $\sum_p \lambda(p) = 1$. For a chain C, the subset of form $\sum_{p \in C} \lambda(p) p$ is called a *simplex*. For a function $g : \Gamma \to \overline{\mathbf{R}}$, the *Lovász extension* $g : K(\Gamma) \to \overline{\mathbf{R}}$ is defined by

$$\bar{g}(x) := \sum_{p \in \Gamma} \lambda(p) g(p) \quad \left(x = \sum_{p \in \Gamma} \lambda(p) p \in K(\Gamma) \right). \tag{9}$$

The space $K(\Gamma)$ is endowed with a natural metric. For an apartment Σ, the subcomplex $K(\Sigma)$ is isomorphic to a simplicial subdivision of $\mathbf{R}^n$ into simplices of vertices

$$z, \quad z + s_{i_1}, \quad z + s_{i_1} + s_{i_2}, \quad \dots, \quad z + s_{i_1} + s_{i_2} + \cdots + s_{i_n}$$

for all even integer vectors z, permutations $(i_1, i_2, \dots, i_n)$ of $\{1, 2, \dots, n\}$ and $s_i \in \{e_i, -e_i\}$ for $i = 1, 2, \dots, n$, where e_i denotes the i-th unit vector. Therefore one can metrize $K(\Gamma)$ so that, for each apartment Σ, the subcomplex $K(\Sigma)$ is an isometric

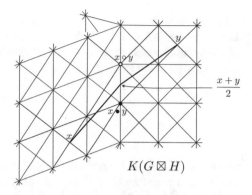

Fig. 3 The space $K(G \boxtimes H)$, which is constructed from $G \boxtimes H$ by filling Euclidean *right* triangle to each chain of length two. The Lovász extension is a piecewise interpolation with respect to this triangulation. Each apartment is isometric to the Euclidean plane, in which geodesics are line segments. The midpoint $(x + y)/2$ of vertices x, y lies on an edge (1-dimensional simplex). Discrete midpoints $x \circ y$ and $x \bullet y$ are obtained by rounding $(x + y)/2$ to the ends of the edge with respect to the partial order

subspace of $K(\Gamma)$ and is isometric to the Euclidean space $(\mathbf{R}^n, l_2)$. Figure 3 illustrates the space $K(\Gamma)$ for twisted tree-grid $\Gamma = G \boxtimes H$. This metric space $K(\Gamma)$ is known as the *standard geometric realization* of Γ; see [1, Chap. 11]. It is known that $K(\Gamma)$ is a *CAT(0) space* (see [7]), and hence uniquely geodesic, i.e., every pair of points can be joined by a unique geodesic (shortest path). The unique geodesic for two points x, y is given as follows. Consider an apartment Σ with $x, y \in K(\Sigma)$, and identify $K(\Sigma)$ with $\mathbf{R}^n$ and x, y with points in $\mathbf{R}^n$. Then the line segment $\{\lambda x + (1 - \lambda)y \mid 0 \leq \lambda \leq 1\}$ in $\mathbf{R}^n \simeq K(\Sigma) \subseteq K(\Gamma)$ is the geodesic between x and y. In particular, a convex function on $K(\Gamma)$ is defined as a function $g : K(\Gamma) \to \overline{\mathbf{R}}$ satisfying

$$\lambda g(x) + (1 - \lambda)g(y) \geq g(\lambda x + (1 - \lambda)y) \quad (x, y \in K(\Gamma), 0 \leq \lambda \leq 1), \quad (10)$$

where point $\lambda x + (1 - \lambda)y$ is considered in the apartment $\mathbf{R}^n \simeq K(\Sigma)$. Notice that the above $(x \bullet y, x \circ y)$ is equal to the unique pair (z, z') with the property that $z \preceq z'$ and $(x + y)/2 = (z + z')/2$ (in $\mathbf{R}^n \simeq K(\Sigma)$). In this setting, the L-convexity is characterized as follows.

Theorem 2.6 ([28]) *Let Γ be a Euclidean building of type C. For $g : \Gamma \to \overline{\mathbf{R}}$, the following conditions are equivalent:*

(1) g *is L-convex.*
(2) The Lovász extension $\bar{g}$ of g is convex.
(3) g *is submodular on I_x and on F_x for every $x \in \mathrm{dom}\, g$, and $\mathrm{dom}\, g$ is chain-connected.*

Here a subset X of vertices in Γ is *chain-connected* if every $p, q \in X$ there is a sequence $p = p_0, p_1, \ldots, p_k = q$ such that $p_i \preceq p_{i+1}$ or $p_i \succeq p_{i+1}$.

Remark 2.7 A Euclidean building Γ consisting of a single apartment is identified with $\mathbf{Z}^n$, and $K(\Gamma)$ is a simplicial subdivision of $\mathbf{R}^n$. Then L-convex functions on Γ as defined here coincide with *UJ-convex functions* by Fujishige [13], where he defined them via the convexity property (2) in Theorem 2.6.

3 Application

In this section, we demonstrate SDA-based algorithm design for two important multiflow problems from which L-convex functions above actually arise. The first one is the minimum-cost multiflow problem, and the second is the node-capacitated multiflow problem. Both problems also arise as (the dual of) LP-relaxations of other important network optimization problems, and are connected to good approximation algorithms. The SDA framework brings fast combinatorial polynomial time algorithms to both problems for which such algorithms had not been known before.

We will use standard notation for directed and undirected networks. For a graph (V, E) and a node subset $X \subseteq V$, let $\delta(X)$ denote the set of edges ij with $|\{i, j\} \cap X| = 1$. In the case where G is directed, let $\delta^+(X)$ and $\delta^-(X)$ denote the sets of edges leaving X and entering X, respectively. Here $\delta(\{i\})$, $\delta^+(\{i\})$ and $\delta^-(\{i\})$ are simply denoted by $\delta(i)$, $\delta^+(i)$ and $\delta^-(i)$, respectively. For a function h on a set V and a subset $X \subseteq V$, let $h(X)$ denote $\sum_{i \in X} h(i)$.

3.1 *Minimum-Cost Multiflow and Terminal Backup*

3.1.1 Problem Formulation

An *undirected network* $N = (V, E, c, S)$ consists of an undirected graph (V, E), an edge-capacity $c : E \to \mathbf{Z}_+$, and a specified set $S \subseteq V$ of nodes, called *terminals*. Let $n := |V|$, $m := |E|$, and $k := |S|$. An *S-path* is a path connecting distinct terminals in S. A *multiflow* is a pair $(\mathscr{P}, f)$ of a set $\mathscr{P}$ of S-paths and a flow-value function $f : \mathscr{P} \to \mathbf{R}_+$ satisfying the capacity constraint:

$$f(e) := \sum \{f(P) \mid P \in \mathscr{P} : P \text{ contains } e\} \leq c(e) \quad (e \in E). \tag{11}$$

A multiflow $(\mathscr{P}, f)$ is simply written as f. The *support* of a multiflow f is the edge-weight defined by $e \mapsto f(e)$. Suppose further that the network N is given a nonnegative edge-cost $a : E \to \mathbf{Z}_+$ and a node-demand $r : S \to \mathbf{Z}_+$ on the terminal set S. The *cost* $a(f)$ of a multiflow f is defined as $\sum_{e \in E} a(e)f(e)$. A multiflow f is said to be *r-feasible* if it satisfies

$$\sum \{f(P) \mid P \in \mathscr{P} : P \text{ connects } s\} \geq r(s) \quad (s \in S). \tag{12}$$

Namely each terminal s is connected to other terminals in at least $r(s)$ flows. The *minimum-cost node-demand multiflow problem (MCMF)* asks to find an r-feasible multiflow of minimum cost.

This problem was recently introduced by Fukunaga [18] as an LP-relaxation of the following network design problem. An edge-weight $u : E \to \mathbf{R}_+$ is said to be r-*feasible* if $0 \le u \le c$ and the network (V, E, u, S) has an $(s, S \setminus \{s\})$-flow of value at least $r(s)$ for each $s \in S$. The latter condition is represented as the following cut-covering constraint:

$$u(\delta(X)) \ge r(s) \quad (s \in S, X \subseteq V : X \cap S = \{s\}). \tag{13}$$

The (capacitated) *terminal backup problem* (TB) asks to find an integer-valued r-feasible edge-weight $u : E \to \mathbf{Z}_+$ of minimum-cost $\sum_{e \in E} a(e)u(e)$. This problem, introduced by Anshelevich and Karagiozova [2], was shown to be polynomially solvable [2, 6] if there is no capacity bound. The complexity of TB for the general capacitated case is not known. The natural fractional relaxation, called the *fractional terminal backup problem (FTB)*, is obtained by relaxing $u : E \to \mathbf{Z}_+$ to $u : E \to \mathbf{R}_+$. In fact, MCMF and FTB are equivalent in the following sense:

Lemma 3.1 ([18])

(1) For an optimal solution f of MCMF, the support of f is an optimal solution of FTB.

(2) For an optimal solution u of FTB, an r-feasible multiflow f in (V, E, u, S) exists, and is optimal to MCMF.

Moreover, half-integrality property holds:

Theorem 3.2 ([18]) *There exist half-integral optimal solutions in FTB, MCMF, and their LP-dual.*

By utilizing this half-integrality, Fukunaga [18] developed a 4/3-approximation algorithm. His algorithm, however, uses the ellipsoid method to obtain a half-integral (extreme) optimal solution in FTB.

Based on the SDA framework of an L-convex function on a tree-grid, the paper [25] developed a combinatorial weakly polynomial time algorithm for MCMF together with a combinatorial implementation of the 4/3-approximation algorithm for TB.

Theorem 3.3 ([25]) *A half-integral optimal solution in MCMF and a 4/3-approximate solution in TB can be obtained in $O(n \log(nAC) \, \mathrm{MF}(kn, km))$ time.*

Here $\mathrm{MF}(n', m')$ denote the time complexity of solving the maximum flow problem on a network of n' nodes and m' edges, and $A := \max\{a(e) \mid e \in E\}$ and $C := \max\{c(e) \mid e \in E\}$.

It should be noted that MCMF generalizes the *minimum-cost maximum free multiflow problem* considered by Karzanov [32, 33]. To this problem, Goldberg and

Karzanov [20] developed a combinatorial weakly polynomial time algorithm. However the analysis of their algorithm is not easy, and the explicit polynomial running
time is not given. The algorithm in Theorem 3.3 is the first combinatorial weakly
polynomial time algorithm having an explicit running time.

3.1.2 Algorithm

Here we outline the algorithm in Theorem 3.3. Let $N = (V, E, c, S)$ be a network,
a an edge cost, and r a demand. For technical simplicity, the cost a is assumed to be
positive integer-valued ≥ 1. First we show that the dual of MCMF can be formulated
as an optimization over the product of *subdivided stars*. For $s \in S$, let G_s be a path
with infinite length and an end vertex v_s of degree one. Consider the disjoint union
$\bigcup_{s \in S} G_s$ and identify all v_s to one vertex O. The resulting graph, called a *subdivided
star*, is denoted by G, and the edge-length is defined as $1/2$ uniformly. Let $d = d_G$
denote the shortest path metric of G with respect to this edge-length.

Suppose that $V = \{1, 2, \ldots, n\}$. A *potential* is a vertex $p = (p_1, p_2, \ldots, p_n)$ in
G^n such that $p_s \in G_s$ for each $s \in S$.

Proposition 3.4 ([25]) *The minimum cost of an r-feasible multiflow is equal to the
maximum of*

$$\sum_{s \in S} r(s) d(p_s, O) - \sum_{ij \in E} c(ij) \max\{d(p_i, p_j) - a(ij), 0\} \qquad (14)$$

over all potentials $p = (p_1, p_2, \ldots, p_n) \in G^n$.

Proof (Sketch) The LP-dual of FTB is given by:

$$\text{Max.} \sum_{s \in S} r(s) \sum_{X : X \cap S = \{s\}} \pi_X - \sum_{e \in E} c(e) \max\{0, \sum_{X : e \in \delta(X)} \pi_X - a(e)\}$$
$$\text{s.t. } \pi_X \geq 0 \ (X \subseteq V : |X \cap S| = 1).$$

By the standard uncrossing argument, one can show that there always exists an
optimal solution π_X such that $\{X \mid \pi_X > 0\}$ is *laminar*, i.e., if $\pi_X, \pi_Y > 0$ it holds
that $X \subseteq Y$, $Y \subseteq X$, or $X \cap Y = \emptyset$. Consider the tree-representation of the laminar
family $\{X \mid \pi_X > 0\}$. Since each X contains exactly one terminal, the corresponding
tree is necessarily a subdivided star $\tilde{G}$ with center O and non-uniform edge-length.
In this representation, each X with $\pi_X > 0$ corresponds to an edge e_X of $\tilde{G}$, and each
node i is associated with a vertex p_i in $\tilde{G}$. The length of each edge e_X is defined
as π_X, and the resulting shortest path metric is denoted by D. Then it holds that
$\sum_{X : X \cap S = \{s\}} \pi_X = D(p_s, O)$ and $\sum_{X : ij \in \delta(X)} \pi_X = D(p_i, p_j)$. By the half-integrality
(Theorem 3.2), we can assume that each π_X is a half-integer. Thus we can subdivide
$\tilde{G}$ to G so that each edge-length of G is $1/2$, and obtain the formulation in (14).

Motivated by this fact, define $\omega : G^n \to \overline{\mathbf{R}}$ by

$$p \mapsto -\sum_{s \in S} r(s)d(p_s, O) + \sum_{ij \in E} c(ij)\max\{d(p_i, p_j) - a(ij), 0\} + I(p),$$

where I is the indicator function of the set of all potentials, i.e., $I(p) := 0$ if p is a potential and ∞ otherwise. The color classes of G are denoted by B and W with $O \in B$, and G is oriented zigzagly. In this setting, the objective function ω of the dual of MCMF is an L-convex function on tree-grid G^n:

Proposition 3.5 ([25]) *The function ω is L-convex on G^n.*

In the following, we show that the steepest descent algorithm for ω is efficiently implementable with a maximum flow algorithm. An outline is as follows:

- The optimality of a potential p is equivalent to the feasibility of a circulation problem on directed network D_p associated with p (Lemma 3.6), where a half-integral optimal multiflow is recovered from an integral circulation (Lemma 3.7).
- If p is not optimal, then a certificate (= violating cut) of the infeasibility yields a steepest direction p' at p (Lemma 3.8).

This algorithm may be viewed as a multiflow version of the dual algorithm [22] on minimum-cost flow problem; see also [49] for a DCA interpretation of the dual algorithm.

Let $p \in G^n$ be a potential and $f : \mathscr{P} \to \mathbf{R}_+$ an r-feasible multiflow, where we can assume that f is positive-valued. Considering $\sum_{e \in E} a(e)f(e) - (-\omega(p))(\geq 0)$, we obtain the complementary slackness condition: Both p and f are optimal if and only if

$$f(e) = 0 \qquad (e = ij \in E : d(p_i, p_j) < a(ij)), \qquad (15)$$

$$f(e) = c(e) \qquad (e = ij \in E : d(p_i, p_j) > a(ij)), \qquad (16)$$

$$\sum_{P \in \mathscr{P}} \{f(P) \mid P \text{ connects } s\} = r(s) \qquad (s \in S : p_s \neq O), \qquad (17)$$

$$\sum_{k=1,\dots,\ell} d(p_{i_{k-1}}, p_{i_k}) = d(p_s, p_t) \quad (P = (s = i_0, i_1, \dots, i_\ell = t) \in \mathscr{P}). \quad (18)$$

The first three conditions are essentially the same as the kilter condition in the (standard) minimum cost flow problem. The fourth one is characteristic of multiflow problems, which says that an optimal multiflow f induces a collection of geodesics in G via embedding $i \mapsto p_i$ for an optimal potential p.

Observe that these conditions, except the fourth one, are imposed on the support of f rather than multiflow f itself. The fourth condition can also be represented by a support condition on an extended (bidirected) network which we construct below; see Fig. 4. An optimal multiflow will be recovered from a fractional bidirected flow on this network.

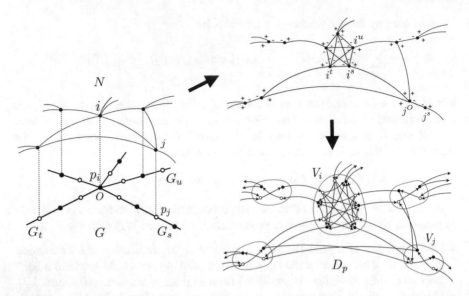

Fig. 4 Construction of extended networks. Each node i is split to a clique of size $|S|$ ($p_i = O$) or size two ($p_i \neq O$), where the edges in these cliques are bidirected edges of sign $(-, -)$. In the resulting bidirected network, drawn in the *upper right*, the path-decomposition of a bidirected flow gives rise to a multiflow flowing in G geodesically. The network D_p, drawn in the *lower right*, is a directed network equivalent to the bidirected network

Let $E_=$ and $E_>$ denote the sets of edges ij with $d(p_i, p_j) = a(ij)$ and with $d(p_i, p_j) > a(ij)$, respectively. Remove all other edges (by (15)). For each nonterminal node i with $p_i = O$, replace i by $|S|$ nodes i^s ($s \in S$) and add new edges $i^s i^t$ for distinct $s, t \in S$. Other node i is a terminal s or a nonterminal node with $p_i \in G_s \setminus \{O\}$. Replace each such node i by two nodes i^s and i^O, and add new edge $i^s i^O$. The node i^s for terminal $i = s$ is also denoted by s. The set of added edges is denoted by E_-. For each edge $e = ij \in E_= \cup E_>$, replace ij by $i^O j^s$ if $p_i, p_j \in G_s$ and $d(p_i, O) > d(p_j, O)$, and replace ij by $i^O j^O$ if $p_i \in G_s \setminus \{O\}$ and $p_j \in G_t \setminus \{O\}$ for distinct $s, t \in S$.

An edge-weight $\psi : E_= \cup E_> \cup E_- \to \mathbf{R}$ is called a *p-feasible support* if

$$0 \le \psi(e) \le c(e) \quad (e \in E_=), \tag{19}$$

$$\psi(e) = c(e) \quad (e \in E_>), \tag{20}$$

$$\psi(e) \le 0 \quad (e \in E_-), \tag{21}$$

$$-\psi(\delta(s)) \ge r(s) \quad (s \in S, p_s = O), \tag{22}$$

$$-\psi(\delta(s)) = r(s) \quad (s \in S, p_s \neq O), \tag{23}$$

$$\psi(\delta(u)) = 0 \quad \text{(each nonterminal node } u\text{)}. \tag{24}$$

One can see from an alternating-path argument that any p-feasible support ψ is represented as a weighted sum $\sum_{P \in \mathscr{P}} f(P) \chi^P$ for a set $\mathscr{P}$ of S-paths with nonnegative coefficients $f : \mathscr{P} \to \mathbf{R}_+$, where each P is a path alternately using edges in E_- and edges in $E_= \cup E_>$, and χ^P is defined by $\chi^P(e) := -1$ for edge e in P with $e \in E_-$, $\chi^P(e) := 1$ for other edge e in P, and zero for edges not in P. Contracting all edges in E_- for all paths $P \in \mathscr{P}$, we obtain an r-feasible multiflow f^ψ in N, where $f^\psi(e) \leq c(e)$ and (16) are guaranteed by (19) and (20), and the r-feasibility and (17) are guaranteed by (22) and (23). Also, by construction, each path in $\mathscr{P}$ induces a local geodesic in G by $i \mapsto p_i$, which must be a global geodesic since G is a tree. This implies (18),

Lemma 3.6 ([25])

(1) A potential p is optimal if and only if a p-feasible support ψ exists.
(2) For any p-feasible support ψ, the multiflow f^ψ is optimal to MCMF.

Thus, by solving inequalities (19)–(24), we obtain an optimal multiflow or know the nonoptimality of p. Observe that this problem is a fractional bidirected flow problem, and reduces to the following circulation problem. Replace each node u by two nodes u^+ and u^-. Replace each edge $e = uv \in E_-$ by two directed edges $e^+ = u^+v^-$ and $e^- = v^+u^-$ with lower capacity $\underline{c}(e^+) = \underline{c}(e^-) := 0$ and upper capacity $\overline{c}(e^+) = \overline{c}(e^-) := \infty$. Replace each edge $e = uv \in E_= \cup E_>$ by two directed edges $e^+ = u^-v^+$ and $e^- = v^-u^+$, where $\overline{c}(e^+) = \overline{c}(e^-) := c(e)$, and $\underline{c}(e^+) = \underline{c}(e^-) :=$ 0 if $e \in E_=$ and $\underline{c}(e^+) = \underline{c}(e^-) := c(e)$ if $e \in E_>$. For each terminal $s \in S$, add edge s^-s^+, where $\underline{c}(s^-s^+) := r(s)$ and $\overline{c}(s^-s^+) := \infty$ if $p_s = O$ and $\underline{c}(s^-s^+) =$ $\overline{c}(s^-s^+) := r(s)$ if $p_s \neq O$. Let D_p denote the resulting network, which is a variant of the *double covering network* in the minimum cost multiflow problem [32, 33].

A *circulation* is an edge-weight φ on this network D_p satisfying $\underline{c}(e) \leq \varphi(e) \leq$ $\overline{c}(e)$ for each edge e, and $\varphi(\delta^+(u)) = \varphi(\delta^-(u))$ for each node u. From a circulation φ in D_p, a p-feasible support ψ is obtained by

$$\psi(e) := (\varphi(e^+) + \varphi(e^-))/2 \quad (e \in E_= \cup E_> \cup E_-). \tag{25}$$

It is well-known that a circulation, if it exists, is obtained by solving one maximum flow problem. Thus we have:

Lemma 3.7 ([25]) *From an optimal potential, a half-integral optimal multiflow is obtained in $O(\mathrm{MF}(kn, m + k^2n)$ time.*

Next we analyze the case where a circulation does not exist. By Hoffman's circulation theorem, a circulation exists in D_p if and only if

$$\kappa(X) := \underline{c}(\delta^-(X)) - \overline{c}(\delta^+(X))$$

is nonpositive for every node subset X. A node subset X is said to be a *violating cut* if $\kappa(X)$ is positive, and is said to be *maximum* if $\kappa(X)$ is maximum among all node-subsets.

From a maximum violating cut, a steepest direction of ω at p is obtained as follows. For (original) node $i \in V$, let V_i denote the set of nodes in D_p replacing

i in this reduction process; see Fig. 4. Let V_i^+ and V_i^- denote the sets of nodes in V_i having $+$ label and $-$ label, respectively. A node subset X is said to be *movable* if $X \cap V_i = \emptyset$ or $\{u^+\} \cup V_i^- \setminus \{u^-\}$ for some $u^+ \in V_i^+$. For a movable cut X, the potential p^X is defined by

$$(p^X)_i = \begin{cases} \text{the neighbor of } p_i \text{ closer to } O & \text{if } X \cap V_i^+ = \{i^{O+}\}, \\ \text{the neighbor of } p_i \text{ in } G_s \text{ away from } O & \text{if } X \cap V_i^+ = \{i^{s+}\}, \quad (26) \\ p_i & \text{if } X \cap V_i = \emptyset. \end{cases}$$

See Fig. 5 for an intuition of p^X. Let $V_I := \bigcup_{i:p_i \in W} V_i$ and $V_F := \bigcup_{i:p_i \in B} V_i$. Since a is integer-valued, edges between V_I and V_F have the same upper and lower capacity. This implies $\kappa(X) = \kappa(X \cap V_I) + \kappa(X \cap V_F)$. Thus, if X is violating, then $X \cap V_I$ or $X \cap V_F$ is violating, and actually gives a steepest direction as follows.

Lemma 3.8 ([25]) *Let p be a nonoptimal potential. For a minimal maximum violating cut X, both $X \cap V_I$ and $X \cap V_F$ are movable. Moreover, $p^{X \cap V_I}$ is a minimizer of ω over I_p and $p^{X \cap V_F}$ is a minimizer of ω over F_p.*

Now SDA is specialized to MCMF as follows.

Steepest Descent Algorithm for MCMF
Step 0:   Let $p := (O, O, \ldots, O)$.
Step 1:   Construct network D_p.
Step 2:   If a circulation φ exists in D_p, then obtain a p-feasible support ψ by (25), and an optimal multiflow f^ψ via the path decomposition; stop.
Step 3:   For a minimal maximum violating cut X, choose $p' \in \{p^{X \cap V_I}, p^{X \cap V_F}\}$ with $\omega(p') = \min\{\omega(p^{X \cap V_I}), \omega(p^{X \cap V_F})\}$, let $p := p'$, and go to step 1.

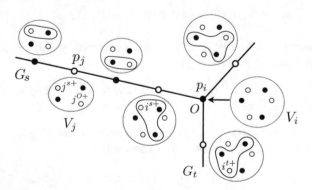

Fig. 5 The correspondence between movable cuts in D_p and neighbors of p. There is a one-to-one correspondence between $I_{p_i} \cup F_{p_i}$ and $\{X \cap V_i \mid X\colon \text{movable cut}\}$, where $X \cap V_i$ is surrounded by a *closed curve*, and nodes with $+$ label and $-$ label are represented by *white* and *black* points, respectively. By a movable cut X, a potential p can be moved to another potential p^X for which $(p^X)_i \in I_{p_i} \cup F_{p_i}$ $(i \in V)$

Proposition 3.9 ([25]) *The above algorithm runs in* $O(nA \cdot \text{MF}(kn, m + k^2 n))$ *time.*

Proof (Sketch) By Theorem 2.5, it suffices to show $\max_i d(O, p_i) = O(nA)$ for some optimal potential p. Let p be an arbitrary optimal potential. Suppose that there is a node i^* such that $p_{i^*} \in G_s$ and $d(p_{i^*}, O) > nA$. Then there is a subpath P of G_s that has $2A$ edges and no node i with $p_i \in P$. Let $U(\ni i^*)$ be the set of nodes i such that p_i is beyond P (on the opposite side to O). For each $i \in U$, replace p_i by its neighbor closer to O. Then one can see that ω does not increase. By repeating this process, we obtain an optimal potential as required. □

This algorithm can be improved to a polynomial time algorithm by a *domain scaling* technique. For a scaling parameter $\ell = -1, 0, 1, 2, \ldots, \lceil \log nA \rceil$, let G_ℓ denote the graph on the subset of vertices x of G with $d(x, O) \in 2^\ell \mathbf{Z}$, where an edge exists between $x, y \in G_\ell$ if and only if $d(x, y) = 2^\ell$. By modifying the restriction of ω to $G_\ell{}^n$, we can define $\omega_\ell : G_\ell{}^n \to \overline{\mathbf{R}}$ with the following properties:

- ω_ℓ is L-convex on $G_\ell{}^n$.
- $\omega_{-1} = \omega$.
- If x_ℓ^* is a minimizer of ω_ℓ over $G_\ell \subseteq G_{\ell-1}$, then $d_\Delta(x_\ell^*, \text{opt}(\omega_{\ell-1})) = O(n)$, where d_Δ is defined for $G_{\ell-1}{}^n$ (with unit edge-length).

The key is the third property, which comes from a *proximity theorem* of L-convex functions [25]. By these properties, a minimizer of ω can be found by calling SDA $\lceil \log nA \rceil$ times, in which x_ℓ^* is obtained in $O(n)$ iterations in each scaling phase. To solve a local k-submodular minimization problem for ω_ℓ, we use a network construction in [31], different from D_p. Then we obtain the algorithm in Theorem 3.3.

3.2 Node-Capacitated Multiflow and Node-Multiway Cut

3.2.1 Problem Formulation

Suppose that the network $N = (V, E, b, S)$ has a node-capacity $b : V \setminus S \to \mathbf{R}_+$ instead of edge-capacity c, where a multiflow $f : \mathscr{P} \to \mathbf{R}_+$ should satisfy the node-capacity constraint:

$$\sum \{ f(P) \mid P \in \mathscr{P} : P \text{ contains node } i \} \le b(i) \quad (i \in V \setminus S). \tag{27}$$

Let $n := |V|$, $m := |E|$, and $k := |S|$ as before. The *node-capacitated maximum multiflow problem (NMF)* asks to a find a multiflow $f : \mathscr{P} \to \mathbf{R}_+$ of the maximum total flow-value $\sum_{P \in \mathscr{P}} f(P)$. This problem first appeared in the work of Garg, Vazirani, and Yannakakis [19] on the node-multiway cut problem. A *node-multiway cut* is a node subset $X \subseteq V \setminus S$ such that every S-path meets X. The *minimum node-multiway cut problem (NMC)* is the problem of finding a node-multiway cut X of minimum capacity $\sum_{i \in X} b(i)$. The two problems NMF and NMC are closely related. Indeed, consider the LP-dual of NMF, which is given by

$$\text{Min.} \quad \sum_{i \in V \setminus S} b(i)w(i)$$

$$\text{s.t.} \quad \sum \{w(i) \mid i \in V \setminus S : P \text{ containsnode } i\} \geq 1 \quad (P : S\text{-path}),$$

$$w(i) \geq 0 \quad (i \in V \setminus S).$$

Restricting w to be 0-1 valued, we obtain an IP formulation of NMC. Garg, Vazirani, and Yannakakis [19] proved the half-integrality of this LP, and showed a 2-approximation algorithm for NMC by rounding a half-integral solution; see also [53]. The half-integrality of the primal problem NMF was shown by Pap [47, 48]; this result is used to solve the integer version of NMF in strongly polynomial time. These half-integral optimal solutions are obtained by the ellipsoid method in strongly polynomial time.

It is a natural challenge to develop an ellipsoid-free algorithm. Babenko [3] developed a combinatorial polynomial time algorithm for NMF in the case of unit capacity. For the general case of capacity, Babenko and Karzanov [4] developed a combinatorial weakly polynomial time algorithm for NMF. As an application of L-convex functions on a twisted tree-grid, the paper [26] developed the first strongly polynomial time combinatorial algorithm:

Theorem 3.10 ([26]) *A half-integral optimal multiflow for NMF, a half-integral optimal solution for its LP-dual, and a 2-approximate solution for NMC can be obtained in $O(m(\log k)\text{MSF}(n, m, 1))$ time.*

The algorithm uses a submodular flow algorithm as a subroutine. Let $\text{MSF}(n, m, \gamma)$ denote the time complexity of solving the *maximum submodular flow problem* on a network of n nodes and m edges, where γ is the time complexity of computing the *exchange capacity* of the defining submodular set function. We briefly summarize the submodular flow problem; see [11, 12] for detail. A *submodular set function* on a set V is a function $h : 2^V \to \overline{\mathbf{R}}$ satisfying

$$h(X) + h(Y) \geq h(X \cap Y) + h(X \cup Y) \quad (X, Y \subseteq V).$$

Let $N = (V, A, \underline{c}, \overline{c})$ be a directed network with lower and upper capacities $\underline{c}, \overline{c} : A \to \mathbf{R}$, and let $h : 2^V \to \mathbf{R}$ be a submodular set function on V with $h(\emptyset) = h(V) = 0$. A *feasible flow* with respect to h is a function $\varphi : A \to \mathbf{R}$ satisfying

$$\underline{c}(e) \leq \varphi(e) \leq \overline{c}(e) \quad (e \in A),$$
$$\varphi(\delta^-(X)) - \varphi(\delta^+(X)) \leq h(X) \quad (X \subseteq V).$$

For a feasible flow φ and a pair of nodes i, j, the *exchange capacity* is defined as the minimum of

$$h(X) - \varphi(\delta^-(X)) + \varphi(\delta^+(X)) \quad (\geq 0)$$

over all $X \subseteq V$ with $i \in X \not\ni j$.

The maximum submodular flow problem (MSF) asks to find a feasible flow φ having maximum $\varphi(e)$ for a fixed edge e. This problem obviously generalizes the maximum flow problem. Generalizing existing maximum flow algorithms, several combinatorial algorithms for MSF have been proposed; see [14] for survey. These algorithms assume an oracle of computing the exchange capacity (to construct the residual network). The current fastest algorithm for MSF is the pre-flow push algorithm by Fujishige-Zhang [17], where the time complexity is $O(n^3\gamma)$. Thus, by using their algorithm, the algorithm in Theorem 3.10 runs in $O(mn^3 \log k)$ time.

3.2.2 Algorithm

Let $N = (V, E, b, S)$ be a network. For several technical reasons, instead of NMF, we deal with a *perturbed* problem. Consider a small uniform edge-cost on E. It is clear that the objective function of NMF may be replaced by $\sum_{P \in \mathscr{P}} Mf(P) - \sum_{e \in E} 2f(e)$ for large $M > 0$. We further purturbe M according to terminals which P connects. Let Σ be a tree such that each non-leaf vertex has degree 3, leaves are u_s ($s \in S$), and the diameter is at most $\lceil \log k \rceil$. For each $s \in S$, consider an infinite path P_s with one end vertex u'_s, and glue Σ and P_s by identifying u_s and u'_s. The resulting tree is denoted by G, where the edge-length is defined as 1 uniformly, and the path-metric on G is denoted by d. Let v_s denote the vertex in P_s with $d(u'_s, v_s) = (2|E| + 1)\lceil \log k \rceil$. The perturbed problem PNMM is to maximize

$$\sum_{P \in \mathscr{P}} d(v_{s_P}, v_{t_P}) f(P) - \sum_{e \in E} 2f(e)$$

over all multiflows $f : \mathscr{P} \to \mathbf{R}_+$, where s_P and t_P denote the ends of an S-path P.

Lemma 3.11 ([26]) *Any optimal multiflow for PNMF is optimal to NMF.*

Next we explain a combinatorial duality of PNMM, which was earlier obtained by [24] for more general setting. Consider the *edge-subdivision* G^* of G, which is obtained from G by replacing each edge pq by a series of two edges pv_{pq} and $v_{pq}q$ with a new vertex v_{pq}. The edge-length of G^* is defined as $1/2$ uniformly, where G is naturally regarded as an isometric subspace of G^* (as a metric space). Let $\mathbf{Z}^* := \{z/2 \mid z \in \mathbf{Z}\}$ denote the set of half-integers. Suppose that $V = \{1, 2, \ldots, n\}$. Consider the product $(G^* \times \mathbf{Z}^*)^n$. An element $(p, r) = ((p_1, r_1), (p_2, r_2), \ldots, (p_n, r_n)) \in (G^* \times \mathbf{Z}^*)^n = (G^*)^n \times (\mathbf{Z}^*)^n$ is called a *potential* if

$$r_i \geq 0 \qquad (i \in V),$$
$$d(p_i, p_j) - r_i - r_j \leq 2 \qquad (ij \in E),$$
$$(p_s, r_s) = (v_s, 0) \qquad (s \in S).$$

and each (p_i, r_i) belongs to $G \times \mathbf{Z}$ or $(G^* \setminus G) \times (\mathbf{Z}^* \setminus \mathbf{Z})$. Corresponding to Proposition 3.4, the following holds:

Proposition 3.12 ([24]) *The optimal value of PNMF is equal to the minimum of* $\sum_{i \in V \setminus S} 2b(i) r_i$ *over all potentials* (p, r).

A vertex $(v_{pq}, z + 1/2)$ in $(G^* \setminus G) \times (\mathbf{Z}^* \setminus \mathbf{Z})$ corresponds to a 4-cycle (p, z), $(p, z + 1)$, $(q, z + 1)$, (q, z) in $G \times \mathbf{Z}$. Thus any potential is viewed as a vertex of a twisted tree-grid $(G \boxtimes \mathbf{Z})^n$. Define $\varpi : (G \boxtimes \mathbf{Z})^n \to \overline{\mathbf{R}}$ by

$$\varpi(p, r) := \sum_{i \in V \setminus S} 2b(i) r_i + I(p, r) \quad ((p, r) \in (G \boxtimes \mathbf{Z})^n),$$

where I is the indicator function of the set of all potentials.

Theorem 3.13 ([26]) ϖ *is L-convex on* $(G \boxtimes \mathbf{Z})^n$.

As in the previous subsection, we are going to apply the SDA framework to ϖ, and show that a steepest direction at $(p, r) \in (G \boxtimes \mathbf{Z})^n$ can be obtained by solving a submodular flow problem on network $D_{p,r}$ associated with (p, r). The argument is parallel to the previous subsection but is technically more complicated.

Each vertex $u \in G$ has two or three neighbors in G, which are denoted by u^α for $\alpha = 1, 2$ or $1, 2, 3$. Accordingly, the neighbors in G^* are denoted by $u^{*\alpha}$, where $u^{*\alpha}$ is the vertex replacing edge uu^α. Let $G_3 \subseteq G$ denote the set of vertices having three neighbors.

Let (p, r) be a potential. Let $E_=$ denote the set of edges ij with $d(p_i, p_j) - r_i - r_j = 2$. Remove other edges. For each nonterminal node i, replace i by two nodes i^1, i^2 if $p_i \notin G_3$ and by three nodes i^1, i^2, i^3 if $p_i \in G_3$. Add new edges $i^\alpha i^\beta$ for distinct α, β. The set of added edges is denoted by E_-. For $ij \in E_=$, replace each edge $ij \in E_=$ by $i^\alpha j^\beta$ for $\alpha, \beta \in \{1, 2, 3\}$ with $d(p_i, p_j) = d(p_i, (p_i)^{*\alpha}) + d((p_i)^{*\alpha}, (p_j)^{*\beta}) + d((p_j)^{*\beta}, p_j)$. Since $d(p_i, p_j) \geq 1$ and G^* is a tree, such neighbors $(p_i)^{*\alpha}$ and $(p_j)^{*\beta}$ are uniquely determined. If $i = s \in S$, the incidence of s is unchanged, i.e., let $i^\alpha = s$ in the replacement. An edge-weight $\psi : E_= \cup E_- \to \mathbf{R}$ is called a (p, r)-*feasible support* if it satisfies

$$\psi(e) \geq 0 \qquad (e \in E_=), \qquad (28)$$
$$\psi(e) \leq 0 \qquad (e \in E_-), \qquad (29)$$
$$\psi(\delta(i^\alpha)) = 0 \qquad (\alpha \in \{1, 2, 3\}), \qquad (30)$$
$$-\psi(i^1 i^2) \leq b(i) \qquad (p_i \notin G_3, r_i = 0), \qquad (31)$$
$$-\psi(i^1 i^2) = b(i) \qquad (p_i \notin G_3, r_i > 0), \qquad (32)$$

$$-\psi(i^1 i^2) - \psi(i^2 i^3) - \psi(i^1 i^3) \le b(i) \quad (p_i \in G_3, r_i = 0), \tag{33}$$
$$-\psi(i^1 i^2) - \psi(i^2 i^3) - \psi(i^1 i^3) = b(i) \quad (p_i \in G_3, r_i > 0) \tag{34}$$

for each edge e and nonterminal node i. By precisely the same argument, a (p, r)-feasible support ψ is decomposed as $\psi = \sum_{P \in \mathscr{P}} f^\psi(P) \chi_P$ for a multiflow $f^\psi : \mathscr{P} \to \mathbf{R}_+$, where the node-capacity constraint (27) follows from (31)–(34). Corresponding to Lemma 3.6, we obtain the following, where the proof goes along the same argument.

Lemma 3.14 ([26])

(1) A potential (p, r) is optimal if and only if a (p, r)-feasible support exists.
(2) For any (p, r)-feasible support ψ, the multiflow f^ψ is optimal to PNMF.

The system of inequalities (28)–(34) is similar to the previous bidirected flow problem (19)–(24). However (33) and (34) are not bidirected flow constraints. In fact, the second constraint (34) reduces to a bidirected flow constraint as follows. Add new vertex i^0, replace edges $i^1 i^2, i^2 i^3, i^1 i^3$ by $i^0 i^1, i^0 i^2, i^0 i^3$, and replace (33) by

$$-\psi(\delta(i^0)) = 2b(i), \quad -\psi(i^0 i^\alpha) \le b(i) \ (\alpha = 1, 2, 3). \tag{35}$$

Then $(\psi(i^1 i^2), \psi(i^2 i^3), \psi(i^1 i^3))$ satisfying (34) is represented as

$$\psi(i^\alpha i^\beta) = (\psi(i^0 i^\alpha) + \psi(i^0 i^\beta) - \psi(i^0 i^\gamma))/2 \quad (\{\alpha, \beta, \gamma\} = \{1, 2, 3\})$$

for $(\psi(i^0 i^1), \psi(i^0 i^2), \psi(i^0 i^3))$ satisfying (35).

We do not know whether (33) admits such a reduction. This makes the problem difficult. Node $i \in V \setminus S$ with $p_i \in G_3, r_i = 0$ is said to be *special*. For a special node i, remove edges $i^1 i^2, i^2 i^3, i^1 i^3$. Then the conditions (30) and (33) for $u = i^1, i^2, i^3$ can be equivalently written as the following condition on degree vector $(\psi(\delta(i^1)), \psi(\delta(i^2)), \psi(\delta(i^3)))$:

$$\psi(\delta(X)) - \psi(\delta(Y)) \le g_i(X, Y) \quad (X, Y \subseteq \{i^1, i^2, i^3\} : X \cap Y = \emptyset), \tag{36}$$

where g_i is a function the set of pairs $X, Y \subseteq \{i^1, i^2, i^3\}$ with $X \cap Y \ne \emptyset$. Such g_i can be chosen as a bisubmodular (set) function, and inequality system (36) says that the degree vector must belong to the *bisubmodular polyhedron* associated with g_i; see [12] for bisubmodular polyhedra. Thus our problem of solving (28)–(32), (35), and (36) is a fractional bidirected flow problem with degrees constrained by a bisubmodular function, which may be called a *bisubmodular flow* problem. However this natural class of the problems has not been well-studied so far. We need a further reduction. As in the previous subsection, for the bidirected network associated with (28)–(32), and (35), we construct the equivalent directed network $D_{p,r}$ with upper capacity $\bar{c}$ and lower capacity $\underline{c}$, where node i^α and edge e are doubled as $i^{\alpha+}, i^{\alpha-}$ and e^+, e^-, respectively. Let V_i denote the set of nodes replacing original $i \in V$, as before. We construct a submodular-flow constraint on $V_i = \{i^{1+}, i^{2+}, i^{3+}, i^{1-}, i^{2-}, i^{3-}\}$ for each special node i so that a (p, r)-feasible support ψ is recovered from any feasible flow φ on $D_{p,r}$ by the following relation:

$$\psi(e) = (\varphi(e^+) + \varphi(e^-))/2. \tag{37}$$

Such a submodular-flow constraint actually exists, and is represented by some submodular set function h_i on V_i; see [26] for the detailed construction of h_i.

Now our problem is to find a flow φ on network $D_{p,r}$ having the following properties:

- For each edge e in $D_{p,r}$, it holds that $\underline{c}(e) \leq \varphi(e) \leq \overline{c}(e)$.
- For each special node i, it holds that

$$\varphi(\delta^-(U)) - \varphi(\delta^+(U)) \leq h_i(U) \quad (U \subseteq V_i). \tag{38}$$

- For other node $i \in V \setminus S$, it holds that

$$\varphi(\delta^-(i^{\alpha\sigma})) - \varphi(\delta^+(i^{\alpha\sigma})) = 0 \quad (\alpha \in \{0, 1, 2, 3\}, \sigma \in \{+, -\}).$$

This is a submodular flow problem, where the defining submodular set function is given by $X \mapsto \sum_{i:\text{special}} h_i(X \cap V_i)$, and its exchange capacity is computed in constant time. Consequently we have the following, where the half-integrality follows from the integrality theorem of submodular flow.

Lemma 3.15 ([26]) *From an optimal potential (p, r), a half-integral optimal multiflow is obtained in $O(\text{MSF}(n, m, 1))$ time.*

By Frank's theorem on the feasibility of submodular flow (see [11, 12]), a feasible flow φ exists if and only if

$$\kappa(X) := \underline{c}(\delta^-(X)) - \overline{c}(\delta^+(X)) - \sum_{i:\text{special}} h_i(X \cap V_i). \tag{39}$$

is nonpositive for every vertex subset X. A *violating cut* is a vertex subset X having positive $\kappa(X)$. By a standard reduction technique, finding a feasible flow or violating cut is reduced to a maximum submodular flow problem, where a minimal violating cut is naturally obtained from the residual graph of a maximum feasible flow.

Again, we can obtain a steepest direction from a minimal violating cut, according to its intersection pattern with each V_i. A vertex subset X is called *movable* if for i with $p_i \in G^* \setminus G$ it holds that $|X \cap V_i| \leq 1$ and for node i with $p_i \in G$ it holds that $X \cap V_i = \emptyset$, V_i^+, V_i^-, $\{i^{\alpha+}\}$, $V_i^- \setminus \{i^{\alpha-}\}$, or $\{i^{\alpha+}\} \cup V_i^- \setminus \{i^{\alpha-}\}$ for some $\alpha \in \{1, 2, 3\}$. For a movable cut X, define $(p, r)^X$ by

Fig. 6 The correspondence between movable cuts in $D_{p,r}$ and neighbors of (p, r). For $(p_i, r_i) \in B$, there is a one-to-one correspondence between F_{p_i} and $\{X \cap V_i \mid X : \text{movable cut}\}$, where the meaning of this figure is the same as in Fig. 5

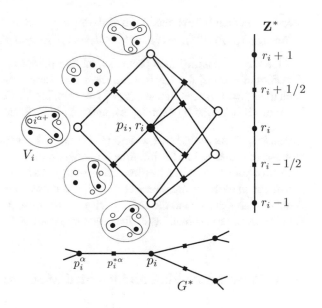

$$(p, r)_i^X := \begin{cases} (p_i, r_i) & \text{if } X \cap V_i = \emptyset, \\ (p_i^{*\alpha}, r_i + 1/2) & \text{if } X \cap V_i = \{i^{\alpha+}\}, \\ (p_i^{*\alpha}, r_i - 1/2) & \text{if } X \cap V_i = V_i^- \setminus \{i^{\alpha-}\}, \\ (p_i^\alpha, r_i) & \text{if } X \cap V_i = \{i^{\alpha+}\} \cup V_i^- \setminus \{i^{\alpha-}\}, \\ (p_i, r_i + 1) & \text{if } X \cap V_i = V_i^+, \\ (p_i, r_i - 1) & \text{if } X \cap V_i = V_i^-. \end{cases} \tag{40}$$

See Fig. 6 for an intuition of $(p, r)^X$. Let V_F be the union of V_i over $i \in V$ with $(p_i, r_i) \in B$ and $\{i^{\alpha+}, i^{\alpha-}\}$ over $i \in V$ and $\alpha \in \{1, 2\}$ with $p_i \in G^* \setminus G$ and $(p_i^{*\alpha}, r_i - 1/2) \in B$. Let V_I be defined by replacing the role of B and W in V_F. Edges between V_I and V_F have the same (finite) lower and upper capacity. If X is violating, then $X \cap V_I$ or $X \cap V_F$ is violating.

Lemma 3.16 ([26]) *Let X be the unique minimal maximum violating cut, and let $\tilde{X}$ be obtained from X by adding V_i^+ for each node $i \in V \setminus S$ with $p_i \in G_3$ and $|X \cap V_i^+| = 2$. Then $\tilde{X} \cap V_I$ and $\tilde{X} \cap V_F$ are movable, one of them is violating, and $(p, r)^{\tilde{X} \cap V_I}$ is a minimizer of ϖ over $I_{p,r}$ and $(p, r)^{\tilde{X} \cap V_F}$ is a minimizer of ϖ over $F_{p,r}$.*

Now the algorithm to find an optimal multiflow is given as follows.

Steepest Descent Algorithm for PNMF

Step 0:   For each terminal $s \in S$, let $(p_s, r_s) := (v_s, 0)$. Choose any vertex v in Σ. For each $i \in V \setminus S$, let $p_i := v$ and $r_i := 2(m + 1)\lceil \log k \rceil$.

Step 1:   Construct $D_{p,r}$ with submodular set function $X \mapsto \sum_{i:\text{special}} h_i(X \cap V_i)$.

Step 2:   If a feasible flow φ exists in $D_{p,r}$, then obtain a (p, r)-feasible support ψ from φ and an optimal multiflow f^ψ via the path decomposition; stop.

Step 3:   For a minimal maximum violating cut X, choose (p', r') with $\varpi(p', r') =$ $\min\{\varpi((p, r)^{\tilde{X} \cap V_I}), \varpi((p, r)^{\tilde{X} \cap V_F})\}$, let $(p, r) := (p', r')$, and go to step 1.

One can see that the initial point is actually a potential. By the argument similar to the proof of Proposition 3.9, one can show that there is an optimal potential (p^*, r^*) such that $r_i^* = O(m \log k)$ and $d(v, p_i^*) = O(m \log k)$. Consequently, the number of iterations is bounded by $O(m \log k)$, and we obtain Theorem 3.10.

Remark 3.17 As seen above, k- and (k, l)-submodular functions arising from localizations of ω and ϖ can be minimized via maximum (submodular) flow. A common feature of both cases is that the domain S_k or $S_{k,l}$ of a k- or (k, l)-submodular function is associated with special intersection patterns between nodes and cuts on which the function-value is equal to the cut-capacity (up to constant). A general framework for such network representations is discussed by Iwamasa [30].

4   L-Convex Function on Oriented Modular Graph

In this section, we explain L-convex functions on oriented modular graphs, introduced in [27, 28]. This class of discrete convex functions is a further generalization of L-convex functions in Sect. 2. The original motivation of our theory comes from the complexity classification of the minimum 0-extension problem. We start by mentioning the motivation and highlight of our theory (Sect. 4.1), and then go into the details (Sects. 4.2 and 4.3).

4.1   Motivation: Minimum 0-Extension Problem

Let us introduce the *minimum 0-extension problem (0-EXT)*, where our formulation is different from but equivalent to the original formulation by Karzanov [34]. An input I consists of number n of variables, undirected graph G, nonnegative weights b_{iv} $(1 \leq i \leq n, v \in G)$ and c_{ij} $(1 \leq i < j \leq n)$. The goal of 0-EXT is to find $x = (x_1, x_2, \ldots, x_n) \in G^n$ that minimizes

$$\sum_{i=1}^{n} \sum_{v \in G} b_{iv} d(x_i, v) + \sum_{1 \leq i < j \leq n} c_{ij} d(x_i, x_j), \tag{41}$$

where $d = d_G$ is the shortest path metric on G. This problem is interpreted as a facility location on graph G. Namely we are going to locate new facilities $1, 2, \ldots, n$ on graph G of cities, where these facilities communicate each other and communicate with all cities, and communication costs are propitional to their distances. The problem is to find a location of minimum communication cost. In facility location theory [50], 0-EXT is known as the *multifacility location problem*. Also 0-EXT is an important special case of the *metric labeling problem* [36], which is a unified label assignment problem arising from computer vision and machine learning. Notice that fundamental combinatorial optimization problems can be formulated as 0-EXT for special underlying graphs. The minimum cut problem is the case of $G = K_2$, and the multiway cut problem is the case of $G = K_m$ $(k \geq 3)$.

In [34], Karzanov addressed the computational complexity of 0-EXT with fixed underlying graph G. This restricted problem class is denoted by 0-EXT[G]. He raised a question: *What are graphs G for which 0-EXT[G] is polynomially solvable?* An easy observation is that 0-EXT[K_m] is in P if $m \leq 2$ and NP-hard otherwise. A classical result [37] in facility location theory is that 0-EXT[G] is in P for a tree G. Consequently, 0-EXT[G] is in P for a tree-product G. It turned out that the tractability of 0-EXT is strongly linked to median and modularity concept of graphs. A *median* of three vertices x_1, x_2, x_3 is a vertex y satisfying

$$d(x_i, x_j) = d(x_i, y) + d(y, x_j) \quad (1 \leq i < j \leq 3).$$

A median is a common point in shortest paths among the three points, may or may not exist, and is not necessarily unique even if it exists. A *median graph* is a connected graph such that every triple of vertices has a *unique* median. Observe that trees and their products are median graphs. Chepoi [9] and Karzanov [34] independently showed that 0-EXT[G] is in P for a median graph G.

A *modular graph* is a further generalization of a median graph, and is defined as a connected graph such that every triple of vertices admits (not necessarily unique) a median. The following hardness result shows that graphs tractable for 0-EXT are necessarily modular.

Theorem 4.1 ([34]) *If G is not orientable modular, then* 0-EXT[G] *is NP-hard.*

Here a (modular) graph is said to be *orientable* if it has an edge-orientation, called an *admissible orientation*, such that every 4-cycle (x_1, x_2, x_3, x_4) is oriented as: $x_1 \to x_2$ if and only if $x_4 \to x_3$. Karzanov [34, 35] showed that 0-EXT[G] is polynomially solvable on special classes of orientable modular graphs.

In [27], we proved the tractability for general orientable modular graphs.

Theorem 4.2 ([27]) *If G is orientable modular, then* 0-EXT[G] *is solvable in polynomial time.*

For proving this result, [27] introduced L-convex functions on oriented modular graphs and submodular functions on modular semilattices, and applied the SDA framework to 0-EXT. An *oriented modular graph* is an orientable modular graph endowed with an admissible orientation. A *modular semilattice* is a semilattice generalization of a modular lattice, introduced by Bandelt, Van De Vel, and Verheul [5]. Recall that a modular lattice L is a lattice such that for every $x, y, z \in L$ with $x \succeq z$ it holds $x \wedge (y \vee z) = (x \wedge y) \vee z$. A modular semilattice is a meet-semilattice L such that every principal ideal is a modular lattice, and for every $x, y, z \in L$ the join

$x \vee y \vee z$ exists provided $x \vee y$, $y \vee z$, and $z \vee x$ exist. These two structures generalize Euclidean buildings of type C and polar spaces, respectively, and are related in the following way.

Proposition 4.3 *(1) A semilattice is modular if and only if its Hasse diagram is oriented modular* [5].
(2) Every principal ideal and filter of an oriented modular graph are modular semilattices [27]. *In particular, every interval is a modular lattice.*
(3) A polar space is a modular semilattice [8].
(4) The Hasse diagram of a Euclidean building of type C is oriented modular [8].

An admissible orientation is acyclic [27], and an oriented modular graph is viewed as (the Hasse diagram of) a poset.

As is expected from these properties and arguments in Sect. 2, an L-convex function on an oriented modular graph is defined so that it behaves submodular on the local structure (principal ideal and filter) of each vertex, which is a modular semilattice. Accordingly, the steepest descent algorithm is well-defined, and correctly obtain a minimizer.

We start with the local theory in the next subsection (Sect. 4.2), where we introduce submodular functions on modular semilattices. Then, in Sect. 4.3, we introduce L-convex functions on oriented modular graphs, and outline the proof of Theorem 4.2.

Remark 4.4 The minimum 0-extension problem 0-EXT[Γ] on a fixed Γ is a particular instance of *finite-valued CSP* with a fixed language. Thapper and Živný [51] established a dichotomy theorem for finite-valued CSPs. The complexity dichotomy in Theorems 4.1 and 4.2 is a special case of their dichotomy theorem, though a characterization of the tractable class of graphs (i.e., orientable modular graphs) seems not to follow directly from their result.

4.2 Submodular Function on Modular Semilattice

A modular semilattice, though not necessarily a lattice, admits an analogue of the join, called the *fractional join*, which is motivated by *fractional polymorphisms* in VCSP [39, 54] and enables us to introduce a submodularity concept.

Let L be a modular semilattice, and let $r : L \to \mathbf{Z}_+$ be the rank function, i.e., $r(p)$ is the length of a maximal chain from the minimum element to p. The fractional join of elements $p, q \in L$ is defined as a formal sum

$$\sum_{u \in E(p,q)} [C(u; p, q)]u$$

of elements $u \in E(p, q) \subseteq L$ with nonnegative coefficients $[C(u; p, q)]$, to be defined soon. Then a function $f : L \to \overline{\mathbf{R}}$ is called *submodular* if it satisfies

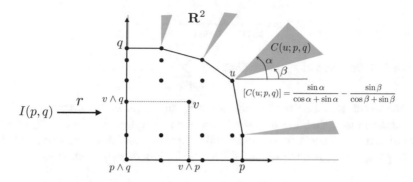

Fig. 7 The construction of the fractional join. By $u \mapsto r(u; p, q)$, the set $I(p, q)$ is mapped to points in $\mathbf{R}_+^2$, where $r(p \wedge q; p, q)$ is the origin, $r(p; p, q)$ and $r(q; p, q)$ are on the coordinate axes. Then Conv $I(p, q)$ is the convex hull of $r(u; p, q)$ over $u \in I(p, q)$. The fractional join is defined as the formal sum of elements mapped to maximal extreme points of Conv $I(p, q)$

$$f(p) + f(q) \geq f(p \wedge q) + \sum_{u \in E(p,q)} [C(u; p, q)] f(u) \quad (p, q \in L).$$

The fractional join of $p, q \in L$ is defined according to the following steps; see Fig. 7 for intuition.

- Let $I(p, q)$ denote the set of all elements $u \in L$ represented as $u = a \vee b$ for some (a, b) with $p \succeq a \succeq p \wedge q \preceq b \preceq q$. This representation is unique, and (a, b) equals $(u \wedge p, u \wedge q)$ [27].
- For $u \in I(p, q)$, let $r(u; p, q)$ be the vector in $\mathbf{R}_+^2$ defined by

$$r(u; p, q) = (r(u \wedge p) - r(p \wedge q), r(u \wedge q) - r(p \wedge q)).$$

- Let Conv $I(p, q) \subseteq \mathbf{R}_+^2$ denote the convex hull of vectors $r(u; p, q)$ over all $u \in I(p, q)$.
- Let $E(p, q)$ be the set of elements u in $I(p, q)$ such that $r(u; p, q)$ is a maximal extreme point of Conv $I(p, q)$. Then $u \mapsto r(u; p, q)$ is injective on $E(p, q)$ [27].
- For $u \in E(p, q)$, let $C(u; p, q)$ denote the nonnegative normal cone at $r(u; p, q)$:

$$C(u; p, q) := \{c \in \mathbf{R}_+^2 \mid \langle c, r(u; p, q) \rangle = \max_{x \in \text{Conv } I(p,q)} \langle c, x \rangle\},$$

where $\langle \cdot, \cdot \rangle$ is the standard inner product.
- For a convex cone $C \subseteq \mathbf{R}_+^2$ represented as

$$C = \{(x, y) \in \mathbf{R}_+^2 \mid y \cos \alpha \leq x \sin \alpha, y \cos \beta \geq x \sin \beta\}$$

for $0 \leq \alpha \leq \beta \leq \pi/2$, define nonnegative value $[C]$ by

$$[C] := \frac{\sin \alpha}{\sin \alpha + \cos \alpha} - \frac{\sin \beta}{\sin \beta + \cos \beta}.$$

- The fractional join of p, q is defined as $\displaystyle\sum_{u \in E(p,q)} [C(u; p, q)]u$.

This weird definition of the submodularity turns out to be appropriate. If L is a modular lattice, then the fractional join is equal to the join $1 \cdot \vee = \vee$, and our definition of submodularity coincides with the usual one. In the case where L is a polar space, it is shown in [28] that the fractional join of p, q is equal to

$$\frac{1}{2}(p \sqcup q) \sqcup q + \frac{1}{2}(p \sqcup q) \sqcup p,$$

and hence a submodular function on L is a function satisfying

$$f(p) + f(q) \geq f(p \wedge q) + \frac{1}{2}f((p \sqcup q) \sqcup q) + \frac{1}{2}f((p \sqcup q) \sqcup p) \quad (p, q \in L). \tag{42}$$

It is not difficult to see that systems of inequalities (42) and (7) define the same class of functions. Thus the submodularity concept in this section is consistent with that in Sect. 2.4.

An important property relevant to 0-EXT is its relation to the distance on L. Let $d : L \times L \to \mathbf{R}$ denote the shortest path metric on the Hasse diagram of L. Then d is also written as

$$d(p, q) = r(p) + r(q) - 2r(p \wedge q) \quad (p, q \in L).$$

Theorem 4.5 ([27]) *Let L be a modular semilattice. Then the distance function d is submodular on $L \times L$.*

Next we consider the minimization of submodular functions on a modular semilattice. The tractability under general setting (i.e., oracle model) is unknown. We consider a restricted situation of *valued constraint satisfaction problem (VCSP)*; see [39, 54] for VCSP. Roughly speaking, VCSP is the minimization problem of a sum of functions with small number of variables. We here consider the following VCSP (*submodular-VCSP on modular semilattice*). An input consists of (finite) modular semilattices $L_1, L_2, \ldots, L_n$ and submodular functions $f_i : L_{i_1} \times L_{i_2} \times \cdots \times L_{i_k} \to \overline{\mathbf{R}}$ with $i = 1, 2, \ldots, m$ and $1 \leq i_1 < i_2 < \cdots < i_k \leq n$, where k is a fixed constant. The goal is to find $p = (p_1, p_2, \ldots, p_n) \in L_1 \times L_2 \times \cdots \times L_n$ to minimize

$$\sum_{i=1}^{m} f_i(p_{i_1}, p_{i_2}, \ldots, p_{i_k}).$$

Each submodular function f_i is given as the table of all function values. Hence the size of the input is $O(nN + mN^k)$ for $N := \max_i |L_i|$.

Kolmogorov, Thapper, and Živńy [39] proved a powerful tractability criterion for general VCSP such that an LP-relaxation (*Basic LP*) exactly solves the VCSP instance. Their criterion involves the existence of a certain submodular-type inequality (*fractional polymorphism*) for the objective functions, and is applicable to our submodular VCSP (thanks to the above weird definition).

Theorem 4.6 ([27]) *Submodular-VCSP on modular semilattice is solvable in polynomial time.*

Remark 4.7 Kuivinen [40, 41] proved a good characterization for general SFM on product $L_1 \times L_2 \times \cdots \times L_n$ of modular lattices L_i with $|L_i|$ fixed. Fujishige, Király, Makino, Takazawa, and Tanigawa [15] proved the oracle-tractability for the case where each L_i is a diamond, i.e., a modular lattice of rank 2.

4.3 L-Convex Function on Oriented Modular Graph

Here we introduce L-convex functions for a slightly restricted subclass of oriented modular graphs; see Remark 4.11 for general case. Recall Proposition 4.3 (2) that every interval of an oriented modular graph Γ is a modular lattice. If every interval of Γ is a *complemented* modular lattice, i.e., every element is a join of rank-1 elements, then Γ is said to be *well-oriented*. Suppose that Γ is a well-oriented modular graph. The L-convexity on Γ is defined along the property (3) in Theorem 2.6, not by discrete midpoint convexity, since we do not know how to define discrete midpoint operations on Γ. Namely an L-convex function on Γ is a function $g : \Gamma \to \mathbf{R}$ such that g is submodular on every principal ideal and filter, and dom g is chain-connected, where the chain-connectivity is similarly defined as in Theorem 2.6. By this definition, the desirable properties hold:

Theorem 4.8 ([27, 28]) *Let g be an L-convex function on Γ. If $x \in$ dom g is not a minimizer of g, then there is $y \in I_x \cup F_x$ with $g(y) < g(x)$.*

Thus the steepest descent algorithm (SDA) is well-defined, and correctly obtains a minimizer of g (if it exists). Moreover the l_∞-iteration bound is also generalized. Let Γ^Δ denote the graph obtained from Γ by adding an edge pq if both $p \wedge q$ and $p \vee q$ exist in Γ, and let $d_\Delta := d_{\Gamma^\Delta}$. Then Theorem 4.9 is generalized as follows.

Theorem 4.9 ([28]) *The number of iterations of SDA applied to L-convex function g and initial point $x \in$ dom g is at most $d_\Delta(x, \mathrm{opt}(g)) + 2$.*

Corresponding to Theorem 4.5, the following holds:

Theorem 4.10 ([27]) *Let G be an oriented modular graph. The distance function d on G is L-convex on $G \times G$.*

We are ready to prove Theorem 4.2. Let G be an orientable modular graph. Endow G with an arbitrary admissible orientation. Then the product G^n of G is oriented modular. It was shown in [27, 28] that the class of L-convex functions is closed under suitable operations such as variable fixing, nonnegative sum, and direct sum. By this fact and Theorem 4.10, the objective function of 0-EXT$[\Gamma]$ is viewed as an L-convex function on G^n. Thus we can apply the SDA framework to 0-EXT$[\Gamma]$, where each local problem is submodular-VCSP on modular semilattice. By Theorem 4.6, a steepest direction at x can be found in polynomial time. By Theorem 4.9, the number of iterations is bounded by the diameter of $(G^n)^\Delta$. Notice that for $x, x' \in G^n$, if $\max_i d(x_i, x'_i) \leq 1$, then x and x' are adjacent in $(G^n)^\Delta$. From this, we see that the diameter of $(G^n)^\Delta$ is not greater than the diameter of G. Thus the minimum 0-extension problem on G is solved in polynomial time.

Remark 4.11 Let us sketch the definition of L-convex function on general oriented modular graph Γ. Consider the poset of all intervals $[p, q]$ such that $[p, q]$ is a complemented modular lattice, where the partial order is the inclusion order. Then the Hasse diagram Γ^* is well-oriented modular [8, 27]. For a function $g : \Gamma \to \overline{\mathbf{R}}$, let $g^* : \Gamma^* \to \overline{\mathbf{R}}$ be defined by $g^*([p, q]) = (g(p) + q(q))/2$. Then an L-convex function on Γ is defined as a function $g : \Gamma \to \overline{\mathbf{R}}$ such that g^* is L-convex on Γ^*. With this definition, desirable properties hold. In particular, the original $\mathrm{L}^\natural$-convex functions coincide with L-convex functions on the product of directed paths, where $\mathbf{Z}$ is identified with an infinite directed path.

Acknowledgements The author thanks Yuni Iwamasa for careful reading, Satoru Fujishige for remarks, and Kazuo Murota for numerous comments improving presentation. The work was partially supported by JSPS KAKENHI Grant Numbers 25280004, 26330023, 26280004, 17K00029.

References

1. P. Abramenko, K.S. Brown, *Buildings-Theory and Applications* (Springer, New York, 2008)
2. E. Anshelevich, A. Karagiozova, Terminal backup, 3D matching, and covering cubic graphs. SIAM J. Comput. **40**, 678–708 (2011)
3. M.A. Babenko, A fast algorithm for the path 2-packing problem. Theory Comput. Syst. **46**, 59–79 (2010)
4. M.A. Babenko, A.V. Karzanov, A scaling algorithm for the maximum node-capacitated multi-flow problem, in *Proceedings of 16th Annual European Symposium on Algorithms (ESA'08)*. LNCS, vol. 5193 (2008), pp. 124–135
5. H.-J. Bandelt, M. van de Vel, E. Verheul, Modular interval spaces. Math. Nachr. **163**, 177–201 (1993)
6. A. Bernáth, Y. Kobayashi, T. Matsuoka, The generalized terminal backup problem. SIAM J. Discret. Math. **29**, 1764–1782 (2015)
7. M.R. Bridson, A. Haefliger, *Metric Spaces of Non-positive Curvature* (Springer, Berlin, 1999)
8. J. Chalopin, V. Chepoi, H. Hirai, D. Osajda, Weakly modular graphs and nonpositive curvature. preprint (2014), arXiv:1409.3892
9. V. Chepoi, A multifacility location problem on median spaces. Discret. Appl. Math. **64**, 1–29 (1996)
10. P. Favati, F. Tardella, Convexity in nonlinear integer programming. Ric. Oper. **53**, 3–44 (1990)

11. A. Frank, *Connections in Combinatorial Optimization* (Oxford University Press, Oxford, 2011)
12. S. Fujishige, *Submodular Functions and Optimization*, 2nd edn. (Elsevier, Amsterdam, 2005)
13. S. Fujishige, Bisubmodular polyhedra, simplicial divisions, and discrete convexity. Discret. Optim. **12**, 115–120 (2014)
14. S. Fujishige, S. Iwata, Algorithms for submodular flows. IEICE Trans. Inf. Syst. **83**, 322–329 (2000)
15. S. Fujishige, T. Király, K. Makino, K. Takazawa, S. Tanigawa, Minimizing submodular functions on diamonds via generalized fractional matroid matchings. EGRES Technical Report (TR-2014-14) (2014)
16. S. Fujishige, K. Murota, Notes on L-/M-convex functions and the separation theorems. Math. Progr. Ser. A **88**, 129–146 (2000)
17. S. Fujishige, X. Zhang, New algorithms for the intersection problem of submodular systems. Japan J. Ind. Appl. Math. **9**, 369–382 (1992)
18. T. Fukunaga, Approximating the generalized terminal backup problem via half-integral multiflow relaxation. SIAM J. Discret. Math. **30**, 777–800 (2016)
19. N. Garg, V.V. Vazirani, M. Yannakakis, Multiway cuts in node weighted graphs. J. Algorithms **50**, 49–61 (2004)
20. A.V. Goldberg, A.V. Karzanov, Scaling methods for finding a maximum free multiflow of minimum cost. Math. Oper. Res. **22**, 90–109 (1997)
21. G. Grätzer, *Lattice Theory: Foundation* (Birkhäuser, Basel, 2011)
22. R. Hassin, The minimum cost flow problem: a unifying approach to dual algorithms and a new tree-search algorithm. Math. Progr. **25**, 228–239 (1983)
23. H. Hirai, Folder complexes and multiflow combinatorial dualities. SIAM J. Discret. Math. **25**, 1119–1143 (2011)
24. H. Hirai, Half-integrality of node-capacitated multiflows and tree-shaped facility locations on trees. Math. Progr. Ser. A **137**, 503–530 (2013)
25. H. Hirai, L-extendable functions and a proximity scaling algorithm for minimum cost multiflow problem. Discret. Optim. **18**, 1–37 (2015)
26. H. Hirai, A dual descent algorithm for node-capacitated multiflow problems and its applications. preprint (2015), arXiv:1508.07065
27. H. Hirai, Discrete convexity and polynomial solvability in minimum 0-extension problems. Math. Progr. Ser. A **155**, 1–55 (2016)
28. H. Hirai, L-convexity on graph structures (2016), arXiv:1610.02469
29. A. Huber, V. Kolmogorov, Towards minimizing k-submodular functions. in *Proceedings of the 2nd International Symposium on Combinatorial Optimization (ISCO'12)*. LNCS, vol. 7422 (Springer, Berlin, 2012), pp. 451–462
30. Y. Iwamasa, On a general framework for network representability in discrete optimization. J. Comb. Optim. (to appear)
31. Y. Iwata, M. Wahlström, Y. Yoshida, Half-integrality, LP-branching and FPT algorithms. SIAM J. Comput. **45**, 1377–1411 (2016)
32. A.V. Karzanov, A minimum cost maximum multiflow problem, in *Combinatorial Methods for Flow Problems*, Institute for System Studies (Moscow, 1979), pp. 138–156 (Russian)
33. A.V. Karzanov, Minimum cost multiflows in undirected networks. Math. Progr. Ser. A **66**, 313–324 (1994)
34. A.V. Karzanov, Minimum 0-extensions of graph metrics. Eur. J. Comb. **19**, 71–101 (1998)
35. A.V. Karzanov, One more well-solved case of the multifacility location problem. Discret. Optim. **1**, 51–66 (2004)
36. J. Kleinberg, É. Tardos, Approximation algorithms for classification problems with pairwise relationships: metric labeling and Markov random fields. J. ACM **49**, 616–639 (2002)
37. A.W.J. Kolen, *Tree Network and Planar Rectilinear Location Theory, CWI Tract 25* (Center for Mathematics and Computer Science, Amsterdam, 1986)
38. V. Kolmogorov, Submodularity on a tree: unifying $L^\natural$-convex and bisubmodular functions. in *Proceedings of the 36th International Symposium on Mathematical Foundations of Computer Science (MFCS'11)*. LNCS, vol. 6907 (Springer, Berlin, 2011), pp. 400–411

39. V. Kolmogorov, J. Thapper, S. Živný, The power of linear programming for general-valued CSPs. SIAM J. Comput. **44**, 1–36 (2015)
40. F. Kuivinen, Algorithms and hardness results for some valued CSPs, dissertation No. 1274, Linköping Studies in Science and Technology, Linköping University, Linköping Sweden (2009)
41. F. Kuivinen, On the complexity of submodular function minimisation on diamonds. Discret. Optim. **8**, 459–477 (2011)
42. L. Lovász, Submodular functions and convexity. in eds. By A. Bachem, M. Grötschel, B. Korte, *Mathematical Programming—The State of the Art* (Springer, Berlin, 1983), pp. 235–257
43. K. Murota, Discrete convex analysis. Math. Progr. **83**, 313–371 (1998)
44. K. Murota, *Discrete Convex Analysis* (SIAM, Philadelphia, 2003)
45. K. Murota, Recent developments in discrete convex analysis, in eds. By W.J. Cook, L. Lovász, J. Vygen, *Research Trends in Combinatorial Optimization* (Springer, Berlin, 2009), pp. 219–260
46. K. Murota, A. Shioura, Exact bounds for steepest descent algorithms of L-convex function minimization. Oper. Res. Lett. **42**, 361–366 (2014)
47. G. Pap, Some new results on node-capacitated packing of A-paths, in *Proceedings of the 39th Annual ACM Symposium on Theory of Computing (STOC'07)* (ACM, New York, 2007), pp. 599–604
48. G. Pap, Strongly polynomial time solvability of integral and half-integral node-capacitate multiflow problems, EGRES Technical Report, TR-2008-12 (2008)
49. A. Shioura, Algorithms for L-convex function minimization: connection between discrete convex analysis and other research fields. J. Oper. Res. Soc. Japan. (to appear)
50. B.C. Tansel, R.L. Francis, T.J. Lowe, Location on networks I. II Manag. Sci. **29**, 498–511 (1983)
51. J. Thapper, S. Živný, The complexity of finite-valued CSPs. J. ACM **63**(37) (2016)
52. J. Tits *Buildings of Spherical Type and Finite BN-pairs*. Lecture Notes in Mathematics, vol. 386 (Springer, New York, 1974)
53. V.V. Vazirani, *Approximation Algorithms* (Springer, Berlin, 2001)
54. S. Živný, *The Complexity of Valued Constraint Satisfaction Problems* (Springer, Heidelberg, 2012)

Parameterized Complexity of the Workflow Satisfiability Problem

D. Cohen, J. Crampton, G. Gutin and M. Wahlström

Abstract The problem of finding an assignment of authorized users to tasks in a workflow in such a way that all business rules are satisfied has been widely studied in recent years. What has come to be known as the workflow satisfiability problem is known to be hard, yet it is important to find algorithms that can solve the problem as efficiently as possible, because it may be necessary to solve the problem multiple times for the same instance of a workflow. Hence, the most recent work in this area has focused on finding optimal fixed-parameter algorithms to solve the problem. In this chapter, we summarize our recent results.

1 Introduction

It is increasingly common for organizations to computerize their business and management processes. The co-ordination of the tasks that comprise a computerized business process is managed by a workflow management system (or business process management system). Typically, the execution of these tasks will be triggered by a human user, or a software agent acting under the control of a human user. As with all multi-user computing systems, it is important to implement some form of access control. In the context of workflow management, this means restricting the execution of each task to some set of authorized users. Existing technologies, such as role-based access control [45], can be used to define and enforce the required authorization policies.

In addition, many workflows require controls on the users that perform groups of tasks. The concept of a Chinese wall, for example, limits the set of tasks that any one user can perform [10], as does separation-of-duty, which is a central part of the role-based access control model [1]. Hence, it is important that workflow management systems implement security controls that enforce authorization rules and business rules, in order to comply with statutory requirements or best practice [4]. It is these "security-aware" workflows that will be the focus of the remainder of this paper.

D. Cohen · J. Crampton · G. Gutin (✉) · M. Wahlström
Royal Holloway, University of London, Egham, Surrey TW20 0EX, UK
e-mail: G.Gutin@rhul.ac.uk

© Springer Nature Singapore Pte Ltd. 2017
T. Fukunaga and K. Kawarabayashi (eds.), *Combinatorial Optimization and Graph Algorithms*, DOI 10.1007/978-981-10-6147-9_5

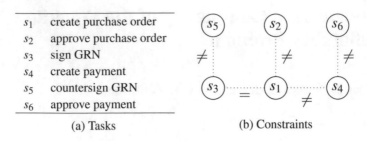

s_1	create purchase order
s_2	approve purchase order
s_3	sign GRN
s_4	create payment
s_5	countersign GRN
s_6	approve payment

(a) Tasks   (b) Constraints

Fig. 1 A simple constrained workflow for purchase order processing

A simple, illustrative example for purchase order processing [21] is shown in Fig. 1. In the first two tasks of the workflow, the purchase order is created and approved (and then dispatched to the supplier). The supplier will submit an invoice for the goods ordered, which is processed by the create payment task. When the supplier delivers the goods, a goods received note (GRN) must be signed and countersigned. Only then may the payment be approved and sent to the supplier. Note that a workflow specification need not be linear: the processing of the GRN and of the invoice can occur in parallel, for example.

In addition to defining the order in which tasks must be performed and of more interest from the security perspective, the workflow specification includes rules to prevent fraudulent use of the purchase order processing system. In our example, these rules restrict the users that can perform pairs of tasks in the workflow: the same user may not sign and countersign the GRN, for example. There may also be a requirement that some tasks are performed by the same user. In our example, the user that raises a purchase order is also required to sign for receipt of the goods.

The rules described above can be encoded using constraints [21], the rules being enforced if and only if the constraints are satisfied. More complex constraints, in which restrictions are placed on the users who execute sets of tasks, can also be defined [2, 27, 47], which encode more complex business requirements. (We describe these constraints in more detail in Sect. 2.1.)

1.1 The Workflow and Constraint Satisfiability Problems

It is apparent that it may be impossible to find an assignment of authorized users to workflow tasks such that all constraints are satisfied. In this case, we say that the workflow specification is *unsatisfiable*. The WORKFLOW SATISFIABILITY PROBLEM (WSP) is known to be NP-hard (see Sect. 2.2), even when the set of constraints only includes constraints that have a relatively simple structure that would arise regularly in practice).

A considerable body of work now exists on the satisfiability of workflow specifications in the presence of constraints [4, 27, 47]. Unsurprisingly, the complexity of WSP is dependent on the constraints defined in the workflow instance. It is convenient, therefore, to consider the complexity of WSP for instances in which the constraints are restricted to those defined by a specific *constraint language*. The instance in Fig. 1, for example, uses a very simple constraint language (and WSP for this language is "easy" in relative terms).

The *Constraint Satisfaction Problem (CSP)* is a general paradigm for expressing, in a declarative format, problems where variables are to be assigned values from some domain. The assignments are constrained by restricting the allowed simultaneous assignments to some sets of variables. This model is useful in many application areas including planning, scheduling, frequency assignment and circuit verification [44].

The CSP is NP-hard, even when only binary not-equals constraints are allowed and the domain has three elements, as we can reduce GRAPH 3- COLORING to the CSP.[1] Hence, a considerable effort has been made to understand the effect of restricting the type of allowed constraints. Recently there has been significant progress towards the completion of this research program and there is now strong evidence to support the algebraic dichotomy conjecture of Bulatov, Jeavons and Krokhin [13], characterising precisely which kinds of constraint language lead to polynomial solvability.

It is worth noting that the WSP is a subclass of the CSP where for each variable s (called a task in WSP terminology) we have an arbitrary unary constraint (called an authorization) that assigns possible values (called users) for s; this is called the conservative CSP. Note, however, that while usually in CSP the number of variables is much larger than the number of values, for the WSP the number of tasks is usually much smaller than the number of users. It is important to remember that for the WSP we do not use the term'constraint' for authorizations and so when we define special types of constraints, we do not extend these types to authorizations, which remain arbitrary.

1.2   Parameterized Complexity Approach to WSP

For many constraint languages, WSP is amenable to the techniques from the field of parameterized complexity. In particular, we are able to find algorithms whose running-times are exponential only in the number k of tasks. Such algorithms are called *fixed-parameter tractable (FPT)* algorithms (with respect to the *parameter* k).[2] The nature of the constraint language determines the exponential term in k. In many practical situations, the number of tasks will be an order of magnitude smaller than the number of users. Thus, FPT algorithms are of considerable value in deciding WSP. Empirical studies have been carried out [17, 40, 41] which show that FPT algorithms for user-independent constraints (for the definition, see Sect. 2.1) outperform the traditionally used SAT4J solver [43]. Since user-independent constraints cover the

---

[1]In fact, NP-hardness result for the WSP is thus a restatement of this well-known result for CSP.
[2]For a formal definition of FPT algorithms, see Sect. 2.3.

majority of practical WSP constraints, the FPT algorithms are not only of theoretical, but also of practical interest.

We provide a brief introduction to parameterized complexity in Sect. 2.3 and describe a number of results that we have obtained for different constraint languages in Sect. 3.

2 Background

We define a *workflow schema* to be a tuple (S, U, A, C), where S is the set of tasks in the workflow, U is the set of users, $A = \{A(s) : s \in S\}$, where $A(s) \subseteq U$ is the *authorization list* for task s, and C is a set of workflow constraints. A *workflow constraint* is a pair $c = (L, \Theta)$, where $L \subseteq S$ and Θ is a set of functions from L to U: L is the *scope* of the constraint; Θ specifies those assignments of elements of U to elements of L that *satisfy* the constraint c. In practice, the elements of Θ are not explicitly defined; in Fig. 2b, we discuss a number of different types of constraints and how the members of Θ are defined implicitly for those constraint types. Following WSP literature, we assume in this chapter that every WSP constraint under consideration can be checked in time polynomial in the numbers of users and tasks.

Given $T \subseteq S$ and $X \subseteq U$, a *plan* is a function $\pi : T \to X$. Given a workflow constraint (L, Θ), $T \subseteq S$ and $X \subseteq U$, a plan $\pi : T \to X$ *satisfies* (L, Θ) if either $L \nsubseteq T$ or $\pi|_L = \theta$ for some $\theta \in \Theta$. In other words, if the range of π includes L then the assignment of users to tasks in L must be one of the satisfying assignments. A plan $\pi : T \to X$ is:

- *eligible* if π satisfies every constraint in C;
- *authorized* if $\pi(s) \in A(s)$ for all $s \in T$;
- *valid* if it is eligible and authorized;
- *complete* if $T = S$.

An algorithm to solve WSP takes a workflow schema (S, U, A, C) as input and outputs a valid, complete plan, if one exists (and null, otherwise).

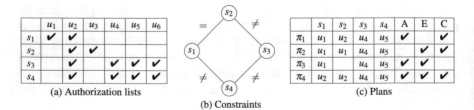

Fig. 2 A simple instance of a workflow schema and plans

Figure 2 illustrates a simple instance of the workflow satisfiability problem. The authorization lists and constraints are defined in Figs. 2a,b respectively. An edge (s, s') labelled with binary relation ρ indicates that the users assigned to s and s' must satisfy ρ. Figure 2c defines four plans π_1, π_2, π_3 and π_4, a blank square indicating that the plan does not assign a task to any user. The table also indicates which of these plans are authorized (A), eligible (E) and complete (C).

- π_1 is a complete plan which is authorized but not eligible, as s_1 and s_2 are assigned to different users.
- π_2 is a complete plan which is eligible but not authorized, as u_1 is not authorized for s_2.
- π_3 is a plan which is authorized and eligible, and therefore valid. However, π_3 is not a complete plan as there is no assignment for s_2.
- π_4 is a complete plan which is eligible and authorized.

Thus π_4 is a valid complete plan, and is therefore a solution to the WSP instance defined in Fig. 2.

2.1   Constraints and Authorization Lists

In this chapter, we consider the complexity of the WSP when the workflow constraint language (which defines the set of permissible workflow constraints) is restricted in some way. We could, for example, only consider constraints in which the scope of each constraint is of size two (as in Fig. 2b), so-called *binary* constraints. In this section we introduce the constraint languages of interest. All of them are useful in that they can be used to encode business rules of practical relevance.

2.1.1   Authorization Lists

An authorization list may be interpreted as a *unary* constraint, where the scope of the constraint is a single task s. Thus $\theta \in \Theta$ may be represented as a pair (s, u) for some $u \in U$. The set Θ may, therefore, be interpreted as a list $A(s) = \{u \in U : (s, u) \in \Theta\}$. In this chapter, we interpret unary constraints as authorization lists and handle such constraints in a different way from other constraints. In particular, authorization lists need not be satisfied for a plan to be eligible.

2.1.2   Binary Constraints

These constraints have been widely studied [21, 47]. Constraints on two tasks, s and s', can (and will) be represented in the form (s, s', ρ), where ρ is a binary relation on U. A plan π satisfies such a constraint if $\pi(s) \, \rho \, \pi(s')$. Writing $=$ to denote the relation $\{(u, u) : u \in U\}$ and $\neq$ to denote the relation $\{(u, v) : u, v \in U, u \neq v\}$, binding-of-duty and separation-of-duty constraints may be represented in the form $(s, s', =)$ and $(s, s', \neq)$, respectively. Generalizations of these constraints that are not restricted to singleton tasks have also been considered [27, 47]: a plan π satisfies a constraint of the form (S', S'', ρ) if there are tasks $s' \in S'$ and $s'' \in S''$ such that $\pi(s') \rho \pi(s'')$.

Crampton et al. [27] considered binary constraints for which ρ is $\sim$ or $\nsim$, where $\sim$ is an *equivalence relation* defined on U. A practical example of such workflow constraints is when the equivalence relation partitions the users into different departments: a binary constraint could, for example, require that two tasks be performed by members of the same department. Crampton et al. [22] also considered binary constraints for which ρ is a *partial order* (defined on U). These constraints are useful when there exists a hierarchy of users (as is usually the case in an organization), defined by relative seniority within the user population.

2.1.3 Counting Constraints

A *tasks-per-user counting constraint* has the form (t_ℓ, t_r, T), where $1 \leqslant t_\ell \leqslant t_r \leqslant k$ and $T \subseteq S$. A plan π satisfies (t_ℓ, t_r, T) if a user performs either no tasks in T or between t_ℓ and t_r tasks. Tasks-per-user counting constraints generalize the cardinality constraints which have been widely studied [6, 39, 45]. Some binary constraints may be represented as tasks-per-user counting constraints: $(s, s', \neq)$ is equivalent to $(1, 1, \{s, s'\})$ (in the sense that the set of satisfying assignments for both constraints is identical); $(s, s', =)$ is equivalent to $(2, 2, \{s, s'\})$.

Two other families of counting constraints are *at-most-r* constraints $(r, Q, \leqslant)$ and *at-least-r* constraints $(r, Q, \geqslant)$, where Q is the scope of the constraint. An at-most-r constraint $(r, Q, \leqslant)$ is satisfied if and only if $|\pi(Q)| \leq r$, and an at-least-r constraint $(r, Q, \geqslant)$ is satisfied if and only if $|\pi(Q)| \geq r$. For $Q = \{s, s'\}$, the at-most-1 constraint is equivalent to $(s, s', =)$ and the at-least-2 constraint is equivalent to $(s, s', \neq)$.

2.1.4 Regular Constraints

For any set of constraints C, a valid plan π induces a partition on the set of tasks: we simply define an equivalence relation on S, where $s \sim s'$ if and only if $\pi(s) = \pi(s')$; the equivalence classes form a partition of S. Clearly the partial plan $\phi : S' \to \{u\}$, where S' is a block in the partition and $\pi(s) = u$ for all $u \in S'$, is eligible (with respect to C). However, the converse does not necessarily hold. That is, a partition $\{S_1, \ldots, S_p\}$ and a collection of eligible plans $\{\pi_1, \ldots, \pi_p\}$, where $\pi_i : S_i \to \{u_i\}$ for some $u_i \in U$, does not necessarily imply that $\pi : S \to U$, where $\pi(s) = u_i$ if and only if $s \in S_i$, is a valid plan.

Informally, some valid plans of the form $\pi : T \to \{u\}$ can never be part of a complete valid plan. Perhaps the simplest example arises for constraints of the form $(s, s', =)$. Then $\pi : \{s\} \to \{u\}$ is an eligible plan (since its domain is a proper subset of the constraint's scope). However, we cannot allocate the remaining tasks to any set of users (distinct from u) because the constraint requires that s and s' be executed by the same user.

Accordingly, we say $T \subseteq S$ is an *eligible set* (with respect to constraint set C) if there exists an eligible complete plan π and a user u such that $\pi^{-1}(u) = T$. Then we say that C is *regular* if for any partition $\{S_1, \ldots, S_p\}$ of S into eligible sets and any set of distinct users $\{u_1, \ldots, u_p\}$, the plan $\pi : S \to \{u_1, \ldots, u_p\}$, such that $\pi(s) = u_i$ if $s \in S_i$, is eligible. Regular constraints were introduced by Crampton et al. [27]; they extend the set of constraints considered by Wang and Li [47]. Crampton et al. [27] showed that the following constraints are regular: $(S', S'', \neq)$; $(S', S'', =)$, where at least one of the sets S', S'' is a singleton; tasks-per-user counting constraints of the form $(1, t, T)$.

2.1.5 User-Independent Constraints

Many business rules are indifferent to the identities of the users that complete a set of tasks; they are only concerned with the relationships between those users. (Per-user authorization lists are the main exception to such rules.) The most obvious example of a user-independent constraint is the requirement that two tasks are performed by different users. A more complex example might require that at least three users are required to complete some sensitive set of tasks.

Thus it is natural to introduce the concept of *user-independent* constraints. A constraint (L, Θ) is user-independent if whenever $\theta \in \Theta$ and $\psi : U \to U$ is a permutation then $\psi \circ \theta$, the composition of the two functions, also belongs to Θ. In other words, user-independent constraints do not distinguish between users. User-independent constraints were introduced by Cohen et al. [18].

Every regular constraint is user-independent, but many user-independent constraints are not regular. Indeed, constraints of the type $(S', S'', =)$ are user-independent, but not necessarily regular [27]. Many counting constraints in the Global Constraint Catalogue [5] are user-independent, but not regular. In particular, the constraint NVALUE, which bounds from above the number of users performing a set of tasks, is user-independent but not regular. Note, however, that constraints of the form $(s', s'', \sim)$ and $(s', s'', \nsim)$ (where $\sim$ is an equivalence relation defined on U), are not user-independent, in general. Finally, authorization lists, when viewed as unary constraints, are not user-independent.

2.1.6 Class-Independent Constraints

Class-independent constraints were introduced in [23] as a generalization of user-independent constraints. Class-independent constraints are of interest when the set of users is partitioned into subsets and we wish some of WSP constraints to take into consideration this partition. For example, consider a university whose staff is partitioned into departments and we wish to request that, in some committee, each user performing a task in $T \subseteq U$ must be from a different department (not just to be a different user).

Informally, given an equivalence relation $\sim$ on U, we say a constraint c is class-independent if, for each plan π that satisfies c, for each step s, and for each permutation π of equivalence classes in $U^{\sim}$, replacing $\pi(s)$ by any user in the class $\phi(\pi(s)^{\sim})$ results in a valid plan. More formally, $c = (T, \Theta)$ is *class-independent for* $\sim$ (or simply $\sim$-*independent*) if for any function θ, $\theta^{\sim} \in \Theta^{\sim}$ implies $\theta \in \Theta$, and for any permutation $\phi : U^{\sim} \rightarrow U^{\sim}$, $\theta^{\sim} \in \Theta^{\sim}$ implies $\phi \circ \theta^{\sim} \in \Theta^{\sim}$. In particular, constraints of the form $(s, s', \sim)$ and $(s, s', \nsim)$ mentioned in Sect. 2.1.2 are $\sim$-independent. Observe that a user-independent constraint is a class-independent constraint for the identity relation.

Modern organizations typically have a hierarchical structure, the organization being successively divided into increasingly fine-grained organizational units. An organization might, for example, have divisions, which are split into departments, which are split into sections, which are split into individuals. A hierarchical structure can be introduced formally as follows. An equivalence relation $\sim$ is said to be a *refinement* of an equivalence relation $\sim'$ if any two elements that are equivalent under $\sim$ are also equivalent under $\sim'$. A sequence $\{\sim_1, \ldots, \sim_r\}$ of equivalence relations is said to be *nested* if $\sim_{q+1}$ is a refinement of $\sim_q$ for each $q \in \{1, \ldots, r-1\}$. We may represent a hierarchical organization by defining a number of nested equivalence relations: $u \sim_1 v$ iff u and v belong to the same organization, $u \sim_2 v$ if and only if u and v belong to the same division, etc.

2.2 Complexity of WSP

For an algorithm that runs on an instance (S, U, A, C) of WSP, we will measure the running time in terms of $n = |U|$, $k = |S|$, and $m = |C|$. (The set A of authorization lists consists of k lists each of size at most n, so we do not need to consider the size of A separately when measuring the running time.) We will say an algorithm runs in polynomial time if it has running time at most $p(n, k, m)$, where $p(n, k, m)$ is polynomial in n, k and m. We assume that all constraints (including the unary authorization constraints) can be checked in polynomial time. This, in turn, means that it takes polynomial time to check whether any plan is authorized and whether it is valid.

We will use the O^* notation, which suppresses polynomial factors. That is, $g(n, k, m) = O^*(h(n, k, m))$ if there exists a polynomial $q(n, k, m)$ such that $g(n, k, m) = O(q(n, k, m)h(n, k, m))$.

One naive algorithm for solving WSP simply enumerates every possible plan and terminates either when a valid complete plan is identified (a YES-instance of the problem) or when all possible plans have been considered (a NO-instance). The running time of this algorithm in the worst case is $O^*(n^k)$ (since the time taken to check the constraints is "absorbed" by the O^* notation). Wang and Li [47] proved that WSP is NP-hard, even when we restrict our attention to binary constraints in which the relation is $\neq$. This is not particularly surprising and the proof follows from a simple reduction of this class of WSP problems to GRAPH COLORING.

2.3 Fixed-Parameter Tractability and WSP

Wang and Li [47] observed that the number of tasks in a workflow is likely to be small relative to the size of the input to the workflow satisfiability problem. This observation led them to study the problem using tools from parameterized complexity and to prove that the problem is fixed-parameter tractable for certain classes of constraints. We now provide a short overview of parameterized complexity, explaining what it means for a problem to be fixed-parameter tractable and summarizing the results obtained by Wang and Li for WSP.

Suppose we have an algorithm that solves an NP-hard problem in time $O(f(k)n^d)$, where n denotes the size of the input to the problem, k is some (small) parameter of the problem, f is some function in k only, and d is some constant (independent of k and n). Then we say the algorithm is a *fixed-parameter tractable* (FPT) algorithm. If a problem can be solved using an FPT algorithm then we say that it is an *FPT problem* and that it belongs to the class FPT. For more information on parameterized complexity, see, e.g., [28, 30].

Wang and Li showed, using an elementary argument, that WSP($\mathcal{L}$) – where $\mathcal{L}$ is the language of constraints of the form $(\{s\}, T, \neq)$ with $s \in S$ and $T \subseteq S$ – is FPT and can be solved in time $O(k^{k+1}N) = O^*(k^{k+1})$, where N is the size of the entire input to the problem [47, Lemma 8]. They also showed that WSP($\mathcal{L}'$) – where $\mathcal{L}'$ is the language of constraints of the form $(\{s\}, T, \rho)$ with $s \in S, T \subseteq S$ and $\rho \in \{\neq, =\}$ – is FPT [47, Theorem 9], using a rather more complex approach: specifically, they constructed an algorithm that runs in time $O^*(k^{k+1}(k-1)^{k2^{k-1}})$. However, WSP is W[1]-hard, in general, which suggests it is not FPT [47, Sect. 4.3]. Thus, it is important to determine which classes of constraint languages are FPT.

3 Fixed-Parameter Tractability Results

In this section, we provide an overview of results from recent research on FPT algorithms for WSP. In particular, we introduce two methods for establishing that WSP($\mathcal{L}$) is FPT for a given constraint language $\mathcal{L}$. The first applies to regular constraints and makes use of their "partition property" and an FPT result for MAX WEIGHTED PARTITION to find valid plans. The second can be applied to user-independent constraints and uses the notion of a pattern to partition all complete plans into classes of complete plans which are equivalent from the point of view of eligibility and then applies a bipartite graph matching algorithm to decide whether a given class of eligible complete plans contains a valid complete plan. The second method can be further extended to class-independent constrains. We also provide lower bounds for complexity of corresponding algorithms which show that the obtained algorithms are optimal in a sense subject to complexity theoretical hypotheses.

3.1 Regular Constraints

Crampton et al. [27] proved the following:

Theorem 1 *Let $W = (S, U, A, C)$ be a workflow specification such that each constraint in C is regular. Then the workflow satisfiability problem for W can be solved in time $O^*(2^k)$.*

The proof of this result reduces an instance of WSP to an instance of the MAX WEIGHTED PARTITION (MWP) problem, which, by a result of Björklund et al. [8], is FPT. The MWP problem takes as input a set S of k elements, an integer $M \geqslant 0$ and n functions $\phi_i : 2^S \to [-M, M]$, $1 \leqslant i \leqslant n$, and outputs a partition of S into n blocks $\{F_1, \ldots, F_n\}$ (some of which may be empty), that maximizes $\sum_{i=1}^n \phi_i(F_i)$. Björklund et al. proved that MWP can be solved in time $O^*(2^k)$ [8, Theorem 4].

Sketch proof of Theorem 1.We construct a binary matrix with n rows (indexed by elements of U) and 2^k columns (indexed by elements of 2^S): every entry in the column labeled by the empty set is defined to be 1; the entry indexed by $u \in U$ and $T \subseteq S$ is defined to be 0 if and only if T is ineligible or there exists $s \in T$ such that $(s, u) \notin A$. In other words, the non-zero matrix entry indexed by u and T indicates that u is authorized for all tasks in T and $\pi : T \to \{u\}$ is eligible and authorized, and thus that u could be allocated to the tasks in T in a valid plan.

The matrix defined above encodes a family of functions $\{\phi_u\}_{u \in U}$, $\phi_u : 2^S \to \{0, 1\}$. We now solve MWP on input S and $\{\phi_u\}_{u \in U}$, which has the effect of determining whether we can "stitch together" different partial plans to form a complete, valid plan. Given that $\phi_u(T) \leqslant 1$, $\sum_{u \in U} \phi_u(T_u) \leqslant n$, with equality if and only if we can find plans $\pi_1, \ldots, \pi_p$, where $\pi_i : T_i \to \{u_i\}$, $\{T_1, \ldots, T_p\}$ is a partition of S and $u_i = u_j$ if and only if $i = j$. Since all constraints are regular, the "union" of such a set of plans is a valid plan and thus the instance is satisfiable if and only if MWP returns a partition having weight n. It is easy to establish that the complexity of the above algorithm is $O^*(2^k)$. This completes the proof sketch.

Interestingly that the complexity $O^*(2^k)$ appears to be optimal: Crampton et al. [27] showed that there is no algorithm of complexity $O^*(2^{o(k)})$ for WSP with regular constraints unless the Exponential Time Hypothesis[3] fails, and Gutin and Wahlström [34] proved that WSP with regular constraints has no algorithms of running time $O^*(2^{ck})$ for any $c < 1$ unless the Strong Exponential Time Hypothesis[4] fails.

---

[3]The Exponential Time Hypothesis [37] states that 3-SAT cannot be solved by an algorithm of running time $O^*(2^{o(n)})$, where n is the number of variables in the input CNF formula.

[4]The   Strong   Exponential   Time   Hypothesis   [36]   states   that

Gutin, Kratsch and Wahlström [33] studied the following question for WSP with regular constraints: When there exists a polynomial-time algorithm for user reduction such that the number of users becomes of polynomial size in k? They have obtained a dichotomy which identifies when it is possible and proved that when it is possible the number of users can be bounded from above by k.

3.2   User-Independent Constraints

WSP with user-independent constrains can be solved by FPT algorithms using a different approach based on the notion of a pattern. A *pattern*[5] is a partition of the set S of tasks into (mutually disjoint) non-empty subsets. The partitions are unordered and their total number is the kth Bell number $\mathscr{B}_k$. A complete plan π has pattern $S_1, \ldots, S_t$, where $S_1 \cup \cdots \cup S_t = S$ and $S_i \cap S_j = \emptyset$ for every $i \neq j$, if there are t distinct users $u_1, \ldots, u_t$ such that $S_i = \pi^{-1}(u_i)$ for every $1 \leq i \leq t$. For a complete plan π let $[\pi]$ denote the set of all complete plans with the same pattern as π. Observe that if all WSP constraints are user-independent then all plans in $[\pi]$ are either eligible or not.

This leads to the following approach for solving WSP with user-independent constraints: for each pattern $\mathscr{P}$ generate a complete plan π with pattern $\mathscr{P}$ and check whether π is eligible or not. If π is eligible, construct a bipartite graph $B(\mathscr{P})$ with bipartition $\mathscr{P}$ and U such that there is an edge between $T \in \mathscr{P}$ and $u \in U$ provided $u \in A(s)$ for each $s \in T$. Observe that a matching M in $B(T)$ which covers all vertices in $\mathscr{P}$ corresponds to an authorized complete plan which maps sets of $\mathscr{P}$ to the matched (by M) users. Since we construct $B(\mathscr{P})$ only if the complete plans with pattern $\mathscr{P}$ are all eligible, any authorized plan among them is a valid complete plan. Since for the kth Bell number we have $\mathscr{B}_k = O(2^{k \log_2 k})$, our approach leads to an algorithm of running time $O^*(2^{k \log_2 k})$.

The most efficient implementations of this approach are based on backtracking and can be found in [40, 41]. In both papers the performance of FPT algorithm is compared to that of SAT4J solver [43]. In many cases, but not all, the FPT algorithm performs much better than SAT4J even if the linear constraints in the SAT4J model use variables which take into consideration the fact that all WSP constraints are user-independent [41].

Interestingly, as in the case of regular constraints, the running time of best FPT algorithms for WSP with user-independent constraints is optimal in the following sense. Cohen et al. [18] showed that WSP with user-independent constraints has

---

(Footnote 4 continued)

$$\lim_{t \to \infty} \inf\{c \geq 0 : \ t\text{-SAT has an algorithm in time } O(2^{cn})\} = 1.$$

[5]This notion was introduced by Cohen et al. [18] in a different but equivalent form; our definition is from [40].

no algorithms of running time $O^*(2^{o(k \log_2 k)})$ unless the Exponential Time Hypothesis fails, and Gutin and Wahlström [34] proved that WSP with user-independent constraints has no algorithms of running time $O^*(2^{ck \log_2 k})$ for any $c < 1$ unless the Strong Exponential Time Hypothesis fails.

3.3 Class-Independent Constraints

Consider a sequence $\{\sim_1, \ldots, \sim_r\}$ of nested equivalence relations, i.e., $\sim_{q+1}$ is a refinement of $\sim_q$ for each $q \in \{1, \ldots, r-1\}$. We may assume without loss of generality that $\sim_r$ is the equality relation; that is, all $\sim_r$-independent constraints are user-independent. This assumption is likely to hold in most organizations, where $\sim_r$ partitions the organization into individuals. If this assumption did not hold then we may always add the equality relation as $\sim_{r+1}$ (which trivially refines $\sim_r$), with no user-independent constraints. This has the effect on our algorithm's running time of increasing r by 1.

Crampton et al. [24] proved that WSP with nested class-independent constraints in r levels and with k steps is FPT with a running time of $O^*(2^{k \log_2(rk)})$. Note that the number r of nested equivalence relations appears only in the logarithm of the running time exponent. This indicates that, in practice, the running time of algorithms for WSP with class-independent constraints is not much larger than those for WSP with user-independent constraints. This is confirmed in [23, 24]. The algorithms in [23, 24] use an extension of the pattern approach, where instead of one pattern for each plan, a sequence of r patterns is used for every plan. For more details, see [23, 24].

4 Characterizing the FPT Languages

In this section we try to understand the limits of the theory presented so far. We have described several classes of constraints for which the associated WSP is fixed parameter tractable. In this section we will call any set of workflow constraints a constraint language. We would like a mathematical characterisation of all constraint languages whose associated WSP is fixed parameter tractable. Armed with such a characterisation we could then decide, for any class of constraints, whether the associated WSP is fixed parameter tractable. Such an analysis of the complexity landscape can be useful in designing constraints for the WSP that will be amenable to practical algorithms.

In this quest, we will define a simple measure, the diversity, for a constraint language. We show that bounded diversity is a sufficient condition for the the associated WSP to be FPT. The regular, user independent and class independent constraints described in this chapter all have bounded diversity. Indeed, diversity is very closely associated with the number of different patterns as defined in Sect. 3.2, and is the basis of the FPT algorithms that we have used to solve the WSP. Unfortunately we are able to demonstrate classes of constraints which have unbounded diversity but whose associated WSP is still FPT.

A workflow schema can be seen, in a natural way, as a classical constraint network in which the tasks are variables, authorisation lists are arbitrary unary constraints, and the domain for every variable is the set of users. Constraint languages that allow arbitrary unary constraints are called conservative. Restricting our attention to certain kinds of workflow constraint corresponds to considering the complexity of classical conservative constraint languages.

In constraint satisfaction, there has been a significant work done to identify classes of constraint networks that can be solved in polynomial time in the number of variables. One effective way to achieve this has been to restrict attention to certain kinds of constraint (constraint languages) [3, 9, 11, 12, 14–16, 19, 20, 29, 42, 46] This work on tractability provides a rich class of constraint languages, so it is natural to ask whether their associated WSP is fixed parameter tractable.

A key question for workflow scheduling is therefore:

Question 2 *For which conservative constraint languages Γ is the class of constraint networks over Γ fixed parameter tractable, parametrised by the number of variables.*

4.1 Equivalent Partial Solutions and a Diversity Measure

Recall that, in our context, the tasks are variables that require the assignment of a user from the domain of users, subject to certain constraints.

Rather than discover a plan for a workflow schema by standard backtracking one variable (task) at a time, we can iterate over subsets of the domain (users). In this approach, we build an ordered domain by ordering the users and find for each initial segment of this domain, which sets of variables can be assigned with elements just from that segment.

Example 3 Consider the n-colouring problem, for the k nodes of a graph, defined as a (binary) constraint network. The variables of the constraint network are the nodes to be coloured. The number of such variables is the parameter k. The domain for each of these variables is the (ordered) set of n colours $\{c_1, \ldots, c_n\}$. Constraints, corresponding to the edges of the graph, restrict some pairs of variables to be different from each other.

To begin we ask which subsets of the variables can be assigned the colour c_1. These are precisely the subsets of the variables that do not include the scope of any constraint.

Now we ask, for each of these subsets, how it can be extended by assigning some new variables the colour c_2, and so on for colour c_3 etc.,

After iteration i, we have a set of partial solutions, each assigning some subset of the variables to the values in the initial segment $[c_1, \ldots, c_i]$ of the domain.

We continue this process until we run out of domain elements, or we have assigned every variable.

It is possible to define an equivalence relation on (partial) plans. The idea is that, during the search procedure defined above we need not remember to which domain element each variable is assigned. This is a special property of colouring, but we can generalise it to other constraint types.

Example 4 When beginning the ith iteration we have assigned colours c_1 to c_{i-1}. One such partial plan π might assign colour c_1 to subset V_1, colour c_2 to subset V_2 etc.,. In this iteration we need to find all possible valid extensions of π to c_i. So we need to find all possible subsets V_i such that $\pi \cup (V_i \mapsto c_i)$ is authorised and satisfies all of the constraints.

In this trivial colouring example, to decide if some V_i is acceptable, we only need to know $X = \cup_{r=1}^{i-1} V_r$ as each colour in $\{c_1, \ldots, c_{i-1}\}$ is equivalent as far as colour c_i is concerned. We can choose any V_i that is disjoint from X where V_i does not include the scope of any constraint.

Hence we only need record a set of subsets of the variables at stage $i - 1$ rather than a set of partial plans assigning variables to the colours $\{c_1, \ldots, c_{i-1}\}$. We have removed the dependence on the domain size, except as it counts the maximum number of iterations, and have clearly reduced the complexity of the algorithm.

This motivates the following definition.

Definition 5 Given a constraint language Γ we say that two functions s_1 and s_2 from X to R are Γ-equivalent for a constraint network with domain D, and variables V, whose constraints are from the language Γ, if the following conditions hold:

- s_1 and s_2 are partial plans.
- for any $s' : (V - X) \to (D - R)$, the extended function $s_1 \cup s'$ is a partial plan if and only if $s_2 \cup s'$ is a partial plan.

We can then define the diversity of a constraint language.

Definition 6 The diversity of a subset of the domain R of a constraint network $P = \langle V, D, C \rangle$ over Γ is the number partial plans, assigning values in R, that are not Γ-equivalent.

For a given domain ordering the diversity of P is the maximum of the diversity of each initial segment of D.

The diversity of P is the minimum of the diversity of P with respect to all orderings of its domain.

The diversity of Γ for domain size n and k variables is the maximum of the diversity of any network over Γ with k variables and domain size n.

Example 7 Let Γ be the language of binary disequality constraints.

We have shown in Examples 3 and 4 that, for any subset X of the variables and subset R of the domain, any two functions from X to R are Γ-equivalent.

It follows that the diversity of Γ for domain size n and k variables is at most 2^k.

Given some effective mechanisms for calculating equivalence of partial plans the following is the kind of theorem that allows us to prove FPT for regular, user-independent and class independent constraints. The symbol O^* ignores polynomial terms.

Theorem 8 [18] *Let Γ have diversity at most w for domain size n and k variables then we can solve a constraint network $\langle V, D, C \rangle$ with $|V| = k, |D| = n$ over Γ in time $O^*(3^k w \log w)$.*

What is more, the nature of the diversity measure means that the (finite) union of languages with bounded diversity also has bounded diversity.

We may therefore hope that the following conjecture is true.

Conjecture 9 A conservative constraint language can only be FPT if its diversity for domain size n and k variables is $O(n^{O(1)} f(k))$.

5   Classical Tractable Conservative Constraint Languages

Many languages have been shown to be FPT, when parametrised by the number of variables in the instance. In this Section we begin to widen the theory for the WSP by considering some of the most widely used, natural conservative constraint languages. We show which of them is FPT, parametrised by the number of variables. The FPT results are direct consequences of the well known polynomial algorithms properties of the constraint languages, so will generalise to other tractable languages. On the other hand, we prove the surprising result that the conservative affine language is W[1]-hard. These results establish the richness of the FPT landscape for the WSP.

Finally, we show that max-closed constraints have a very large diversity which serves to refute Conjecture 9.

Example 10 The ZOA language [20] consists of all constraints of one of three kinds. One-Fan, Two-Fan and Permutation. This classical tractable language is conservative. It is well known that enforcing path consistency (PC) is a decision procedure for the ZOA constraint language. The complexity of enforcing PC is $O(k^3 n^3)$ [35], where k is the number of variables and n is the size of the domain. Hence ZOA is FPT parametrised by the number of variables.

The max-closed language [38] is defined over any ordered domain. If any constraint allows two tuples then it also must allow their componentwise maximum. This classical tractable language is conservative. It is well known that enforcing generalised arc consistency (GAC) is a decision procedure for this language. GAC can be enforced on a network with worst-case time complexity $O(e^2 n^2)$ where e is the number of constraints and n the domain size. [7]. In the case of the WSP, the number of variables is bounded by the parameter k so e is bounded by 2^k. Hence the language of max-closed constraints is FPT parametrised by the number of variables.

The affine language [15] is defined over any finite field (for instance the field $\mathbb{F}_p$ of integers modulo a prime p). Its constraints enforce that the values for variables satisfy a linear equation. We can add all unary constraints to this language to obtain the conservative affine language. We prove below in Theorem 11 that this language is W[1]-hard.

Theorem 11 *The conservative affine language is W[1]-hard, parametrised by the number of variables*

Proof The parametrised clique problem is to determine for a given input graph, whether it has a clique on k vertices. This problem is W[1]-hard [31].

We will give an FPT reduction to the class of constraint networks over the conservative affine language.

Let G be an input graph. We wish to determine if G has a clique on k vertices.

A Sidon set [32] is a set of integers with the property that every pair has a different sum. For all n there exists a Sidon set of size at least $\sqrt{n}(1 - o(1))$.

Suppose that G has r vertices. Choose a Sidon set S with r elements embedded in a domain $\mathbb{F}_p$ of size at most $4r^2$. We will build a constraint network $P_G = \langle W \cup Z, \mathbb{F}_p, C \rangle$ where $|W| = k$ and $|Z| = k^2$. Identify each element of Z with a pair of variables in W. Hence each variable in Z is named with a pair $\langle x, y \rangle$ of variables x and y from W.

Add a unary constraint to each variable in W that forces its value to lie in S.

Each vertex a of G corresponds to a value $s(a)$ in S. Associate with an edge (a, b) of G the value $s(a) + s(b)$ in $\mathbb{F}_p$. Since S is a Sidon no two edges of G are associated with the same value. Add a unary constraint to each variable of Z forcing it to take a value associated with an edge of G.

Now, for each variable $z = \langle x, y \rangle$ in Z, add the affine constraint between x, y and z which forces the value of z to be the sum of the values of the variables x and y.

Suppose that C is a k-clique of G. Assign the variables in W the values associated with the k vertices in C. Now let the variable $z = \langle x, y \rangle$ in Z have the unique value allowed by the constraint on x, y and z. Since C is a clique the values assigned to x and y are associated with the vertices of an edge in G so z has a value satisfying its unary constraint and we have a solution to the constraint network.

Conversely if we have any solution to the constraint network then the vertices of G associated with the values assigned to the variables of W will form a k-clique of G.

So we are done.

To conclude this section we prove that the diversity of the max-closed language is sufficiently large to refute Conjecture 9.

Theorem 12 *The diversity of the max-closed language for domain size n and k variables is at least $O(n^k)$.*

Proof Let $D_n = \{1, \ldots, 2n + 1\}$ and R_n be the following binary relation over D_n:

$$R_n = \bigcup_{x=1}^{2n+1} \{x\} \times \{\max(1, x - n), \ldots, \min(2n + 1, x + n)\} .$$

We first show that R_n is max-closed. Consider any two tuples $t = \langle x, y \rangle$ and $t' = \langle x', y' \rangle$ in R_n. We need that $\max(t, t') \in R_n$. If $x = x'$ or $x < x'$ and $y <= y'$ then $\max(t, t') \in \{t, t'\}$ and there is nothing to prove. So we can assume that $x < x'$ and $y > y'$.

Since $\max(t, t') = \langle x', y \rangle$, we require: $\max(1, x' - n) \leq y \leq \min(2n + 1, x' + n)$.

The tuple $\langle x, y \rangle$ is in R_n so we know that $y \leq \min(2n + 1, x + n)$. By assumption $x < x'$, and so $y \leq \min(2n + 1, x + n) \leq \min(2n + 1, x' + n)$.

The tuple $\langle x', y' \rangle$ is also in R_n, so we know that $\max(1, x' - n) \leq y'$. By assumption $y > y'$, and so $\max(1, x' - n) \leq y' \leq y$ and we are done.

We observe without proof that R_n is symmetric and that each value in $1, \ldots, 2n + 1$ is compatible with a distinct subset of values in R_n.

Consider the constraint network with k pairs of variables $\langle v_i, w_i \rangle, i = 1, \ldots, k$. Each pair is constrained by R_n. For any choice of n values from D_n there are n^k partial assignments to $\{v_1, \ldots, v_k\}$ which are not $\{R_n\}$-equivalent. So the diversity of this constraint network is at least n^k.

It follows that the diversity of the max-closed language for domain size n and k variables is at least $O(n^k)$ and the Conjecture 9 is refuted.   □

The connection between constraint networks and the workflow scheduling problem has only recently been exploited by researchers. It is therefore timely to consider how the work done on the complexity of constraint satisfaction can carry over to derive new results for the WSP, and we have begun that work here. Indeed, by refuting Conjecture 9 we have opened up the FPT landscape for the WSP.

We have shown in this section that work on classical tractable constraint languages is unlikely to be of use in the WSP. However, such languages have been enough to show that bounded diversity is not a characterisation. Even so, bounded diversity has been enough to unify the constraint languages described in this chapter that arise naturally in the workflow context.

6 Concluding Remarks

In this chapter, we have described recent advances in our understanding of the fixed-parameter tractability of the workflow satisfiability problem. These advances include improved running times for instances of the problem using classes of constraints that were known to be FPT; identifying new classes of constraints for which the problem is FPT; and developing FPT algorithms that can solve instances which use different classes of constraints.

Crampton et al. [26] (for a preliminary version, see [25]) introduced the BI-OBJECTIVE WORKFLOW SATISFIABILITY PROBLEM (BO-WSP), which enables one to solve *optimization* problems related to workflows and security policies. In particular, we will be able to compute a "least bad" plan when some components of the security policy may be violated. In general, BO-WSP is intractable from both classical and parameterized complexity point of view (the parameter is the number of steps). Crampton et al. [26] proved that if we restrict our attention to user-independent constraints, then computing a Pareto front of BO-WSP is FPT. This result has important practical consequences, since most constraints in the WSP literature that seem to be of practical interest are user-independent. The proof in [26] is constructive and lead to an algorithm, the implementation of which is described and evaluated in [26].

Also Crampton et al. [26] studied the important question of workflow resiliency and prove new results establishing that the classical decision problems are FPT when restricted to user-independent constraints. Crampton et al. [26] argued that existing models of workflow resiliency are unlikely to be suitable in practice and proposed a richer and more realistic model. They concluded by demonstrating that many questions related to resiliency in the context of this new model may be reduced to instances of BO-WSP.

References

1. American National Standards Institute. in *ANSI INCITS 359-2004 for Role Based Access Control* (2004)
2. A. Armando, S. Ponta, Model checking of security-sensitive business processes, in eds. By P. Degano, J.D. Guttman, *Formal Aspects in Security and Trust*. Lecture Notes in Computer Science, vol. 5983 (Springer, 2009), pp. 66–80
3. L. Barto, M. Kozik, Constraint satisfaction problems solvable by local consistency methods. J. ACM **61**(1), 3 (2014)
4. D.A. Basin, S.J. Burri, G. Karjoth, Obstruction-free authorization enforcement: aligning security with business objectives, in *CSF* (IEEE Computer Society, 2011), pp. 99–113
5. N. Beldiceanu, M. Carlsson, J.-X. Rampon, *Global Constraint Catalog*, 2nd edn. (revision a). Technical Report T2012:03, Swedish Institute of Computer Science (2012)
6. E. Bertino, E. Ferrari, V. Atluri, The specification and enforcement of authorization constraints in workflow management systems. ACM Trans. Inf. Syst. Secur. **2**(1), 65–104 (1999)
7. C. Bessière, J.-C. Régin, R.H.C. Yap, Y. Zhang, An optimal coarse-grained arc consistency algorithm. Artif. Intell. **165**(2), 165–185 (2005)
8. A. Björklund, T. Husfeldt, M. Koivisto, Set partitioning via inclusion-exclusion. SIAM J. Comput. **39**(2), 546–563 (2009)
9. M. Bodirsky, J. Nešetřil, Constraint satisfaction with countable homogeneous templates. J. Logic Comput. **16**, 359–373 (2006)
10. D.F.C. Brewer, M.J. Nash, The Chinese wall security policy, in *IEEE Symposium on Security and Privacy* (IEEE Computer Society, 1989), pp. 206–214
11. A. Bulatov, A dichotomy theorem for constraints on a three-element set, in *Proceedings 43rd IEEE Symposium on Foundations of Computer Science, FOCS'02* (IEEE Computer Society, 2002), pp. 649–658
12. A. Bulatov, Tractable conservative constraint satisfaction problems, in *Proceedings 18th IEEE Symposium on Logic in Computer Science, LICS'03* (Ottawa, Canada, 2003), pp. 321–330
13. A. Bulatov, P. Jeavons, A. Krokhin, Classifying the complexity of constraints using finite algebras. SIAM J. Comput. **34**(3), 720–742 (2005)
14. A.A. Bulatov, On the CSP dichotomy conjecture, in eds. By A.S. Kulikov, N.K. Vereshchagin, *Proceedings of the Computer Science-Theory and Applications-6th International Computer Science Symposium in Russia, CSR 2011, St. Petersburg, Russia, June 14-18, 2011*. Lecture Notes in Computer Science, vol. 6651 (Springer, 2011), pp. 331–344
15. A.A. Bulatov, V. Dalmau, A simple algorithm for Mal'tsev constraints. SIAM J. Comput. **36**(1), 16–27 (2006)
16. A.A. Bulatov, D. Marx, The complexity of global cardinality constraints. in *Proceedings of the 24th Annual IEEE Symposium on Logic in Computer Science, LICS 2009, 11-14 August 2009, Los Angeles, CA, USA* (IEEE Computer Society, 2009), pp. 419–428

17. D. Cohen, J. Crampton, A. Gagarin, G. Gutin, M. Jones, Engineering algorithms for workflow satisfiability problem with user-independent constraints, in eds. By J. Chen, J. Hopcroft, J. Wang, *Frontiers in Algorithmics, FAW 2014*. Lecture Notes in Computer Science, vol. 8497 (Springer, 2014), pp. 48–59

18. D. Cohen, J. Crampton, A. Gagarin, G. Gutin, M. Jones, Iterative plan construction for the workflow satisfiability problem problem. J. Artif. Intell. Res. (JAIR) **51**, 555–577 (2014)

19. D. Cohen, P. Jeavons, P. Jonsson, M. Koubarakis, Building tractable disjunctive constraints. J. ACM **47**, 826–853 (2000)

20. M.C. Cooper, D.A. Cohen, P. Jeavons, Characterising tractable constraints. Artif. Intell. **65**(2), 347–361 (1994)

21. J. Crampton, A reference monitor for workflow systems with constrained task execution, in eds. By E. Ferrari, G.-J. Ahn, *SACMAT* (ACM, 2005), pp. 38–47

22. J. Crampton, R. Crowston, G. Gutin, M. Jones, M.S. Ramanujan, Fixed-parameter tractability of workflow satisfiability in the presence of seniority constraints, in eds. By M.R. Fellows, X. Tan, B. Zhu, *FAW-AAIM*. Lecture Notes in Computer Science, vol. 7924 (Springer, 2013), pp. 198–209

23. J. Crampton, A.V. Gagarin, G. Gutin, M. Jones, On the workflow satisfiability problem with class-independent constraints, in *10th International Symposium on Parameterized and Exact Computation, IPEC 2015*. LIPIcs, vol. 43 (2015), pp. 66–77

24. J. Crampton, A.V. Gagarin, G. Gutin, M. Jones, M.M. Wahlström, On the workflow satisfiability problem with class-independent constraints for hierarchical organizations. ACM Trans. Priv. Secur. **19**(3), 8:1–8:29 (2016)

25. J. Crampton, G. Gutin, D. Karapetyan, Valued workflow satisfiability problem, in *Proceedings of the 20th ACM Symposium on Access Control Models and Technologies (SACMAT 2015)* (2015), pp. 3–13

26. J. Crampton, G. Gutin, D. Karapetyan, R. Watrigant, The bi-objective workflow satisfiability problem and workflow resiliency. J. Comput. Secur. **25**(1), 83–115 (2017)

27. J. Crampton, G. Gutin, A. Yeo, On the parameterized complexity and kernelization of the workflow satisfiability problem. ACM Trans. Inf. Syst. Secur. **16**(1), 4 (2013)

28. M. Cygan, F. Fomin, L. Kowalik, D. Lokshtanov, D. Marx, M. Pilipczuk, M. Pilipczuk, S. Saurabh, *Parameterized Algorithms* (Springer, 2015)

29. V. Dalmau, A new tractable class of constraint satisfaction problems, in *Proceedings 6th International Symposium on Artificial Intelligence and Mathematics* (2000)

30. R.G. Downey, M.R. Fellows, *Fundamentals of Parameterized Complexity* (Springer, New York, 2013)

31. R.G. Downey, M.R. Fellows, *Fundamentals of Parameterized Complexity, Texts in Computer Science* (Springer, 2013)

32. P. Erdős, P. Turán, On a problem of sidon in additive number theory, and on some related problems. J. Lond. Math. Soc. **s1-16**(4), 212–215 (1941)

33. G. Gutin, S. Kratsch, M. Wahlström, Polynomial kernels and user reductions for the workflow satisfiability problem. Algorithmica **75**, 383–402 (2016)

34. G. Gutin, M. Wahlström, Tight lower bounds for the workflow satisfiability problem based on the strong exponential time hypothesis. Inf. Process. Lett. **116**(3), 223–226 (2016)

35. C. Han, C.-H. Lee, Comments on Mohr and Henderson's path consistency algorithm. Artif. Intell. **36**, 125–130 (1988)

36. R. Impagliazzo, R. Paturi, On the complexity of k-sat. J. Comput. Syst. Sci. **62**(2), 367–375 (2001)

37. R. Impagliazzo, R. Paturi, F. Zane, Which problems have strongly exponential complexity? J. Comput. Syst. Sci. **63**(4), 512–530 (2001)

38. P.G. Jeavons, M.C. Cooper, Tractable constraints on ordered domains. Artif. Intell. **79**(2), 327–339 (1995)

39. J. Joshi, E. Bertino, U. Latif, A. Ghafoor, A generalized temporal role-based access control model. IEEE Trans. Knowl. Data Eng. **17**(1), 4–23 (2005)

40. D. Karapetyan, A.V. Gagarin, G. Gutin, Pattern backtracking algorithm for the workflow satisfiability problem with user-independent constraints, in *Proceedings of the Frontiers in Algorithmics-9th International Workshop, FAW 2015, Guilin, China, July 3–5* (2015), pp. 138–149

41. D. Karapetyan, A.J. Parkes, G. Gutin, A. Gagarin, Pattern-based approach to the workflow satisfiability problem with user-independent constraints, in *CoRR* (2016), arxiv:abs/1604.05636

42. A. Krokhin, P. Jeavons, P. Jonsson, Reasoning about temporal relations: the tractable subalgebras of Allen's interval algebra. J. ACM **50**, 591–640 (2003)

43. D. Le Berre, A. Parrain, The SAT4J library, release 2.2. J. Satisf. Bool. Model. Comput. **7**, 59–64 (2010)

44. F. Rossi, P.V. Beek, T. Walsh, *Handbook of Constraint Programming (Foundations of Artificial Intelligence)* (Elsevier Science Inc., New York, NY, USA, 2006)

45. R.S. Sandhu, E.J. Coyne, H.L. Feinstein, C.E. Youman, Role-based access control models. IEEE Comput. **29**(2), 38–47 (1996)

46. T.J. Schaefer, The complexity of satisfiability problems, in *Proceedings 10th ACM Symposium on Theory of Computing, STOC'78* (1978), pp. 16–226

47. Q. Wang, N. Li, Satisfiability and resiliency in workflow authorization systems. ACM Trans. Inf. Syst. Secur. **13**(4), 40 (2010)

Printed in the United States
By Bookmasters